W9-AZD-457

DOES
MEASUREMENT
MEASURE
UP?

DOES
MEASUREMENT
MEASURE
UP?

*How Numbers Reveal and
Conceal the Truth*

John M. Henshaw

THE JOHNS HOPKINS UNIVERSITY PRESS
Baltimore

© 2006 The Johns Hopkins University Press
All rights reserved. Published 2006
Printed in the United States of America on acid-free paper
9 8 7 6 5 4 3 2 1

The Johns Hopkins University Press
2715 North Charles Street
Baltimore, Maryland 21218-4363
www.press.jhu.edu

Library of Congress Cataloging-in-Publication Data

Henshaw, John M.
Does measurement measure up? : how numbers reveal and conceal the truth /
John M. Henshaw.
p. cm.
Includes bibliographical references and index.
ISBN 0-8018-8375-X (acid-free paper)
1. Mensuration. 2. Uncertainty (Information theory) 3. Knowledge, Theory of.
I. Title.
QA465.H46 2006
530.8—dc22   2005027716

A catalog record for this book is available from the British Library.

For Mia

CONTENTS

PREFACE

The British comedian Eddie Izzard launches his stand-up comedy routine in *Glorious* (1997) by apologizing to the audience for the presence of a video crew, with their cameras and lights. To make up for this, Izzard promises, "I'm going to be *extra*-funny tonight—an extra 10% funny." Then he adds, mockingly, "You can't check, can you?" Of course not. It is absurd to think that one could really measure how funny something is. Or is it? Hollywood television and movie producers have developed sophisticated ways to measure how funny something is. Millions of dollars, in fact, ride on their ability to measure "funniness."

But lots of other things depend on measurement. Will your son be admitted to his college of choice? That depends on his high school grade point average and SAT scores—measures of his intellectual ability and academic accomplishments.

Will our government regulate carbon dioxide emissions, perhaps drastically changing our transportation systems and energy infrastructure? That depends on a bewildering jumble of complicated measurements that form the core of the global warming debate—things like atmospheric and oceanic temperatures and CO_2 levels—and on how those measurements are analyzed and interpreted by the experts.

Will Alan Greenspan and his colleagues at the Federal Reserve Board raise or lower interest rates at their next meeting? That depends on how they interpret the trends in various measurements of the national and global economies—things like the gross domestic product, the unemployment rate, and the balance of trade.

Will your alma mater's football team, an undefeated powerhouse, be invited to play for the national championship? That depends on your team's Bowl Championship Series (BCS) rating—a measurement of ability based on wins, losses, and the strength of a team's opponents.

Will a man convicted of rape and murder be released from prison after ten years on death row? That depends on whether measurements of various markers in his DNA match those of bodily fluids left at the crime scene.

Will your doctor prescribe powerful, expensive drugs to help clear your arteries of dangerous deposits? That depends on measurements of both the "good" and "bad" levels of cholesterol in your blood and how those measurements change over time.

The British physicist William Thomson, better known as Lord Kelvin (1824–1907), claimed that without measurement, our knowledge of something—anything—is "meager and unsatisfactory." I have often thought about Kelvin's words over the years. As an engineering student, I began to master my chosen profession through quantification—by measuring things. I learned to think, not "The bridge is really strong" (meager and unsatisfactory knowledge), but rather "The bridge has a safe load limit of 10 tons" (now I really know something about the bridge). That is a simple enough example. But how should I think about the abilities of a colleague, who may be the engineer who designed that bridge? Should I consider that she is highly intelligent, creative, diligent, and tenacious? Would my opinion be altered if I knew that in college she only had a 3.2 grade point average, and that she made a 1250 on the SAT? Is the former meager and unsatisfactory with respect to the latter? Or vice versa?

This book is an exploration of the ways in which knowledge and measurement are interrelated. It is a book about what we measure, why we measure, and how we measure. Measurement, it would seem, is much broader and more pervasive than Lord Kelvin imagined. Kelvin believed that, in order to know something about anything, you have to be able to measure it. The Greek mathematician Philolaus put it even more bluntly in the fifth century BC: "Everything that can be known has a number."

Twenty-five centuries later, it seems as though we live in a world ruled by numbers, from IQ to pain-level measurements to "impact factors" that rate the importance of a particular piece of research. Sorting out which measurements have meaning and which do not is a skill. The sheep in George

Orwell's *Animal Farm* end up bleating, "Four legs good, two legs better." That is a measurement too—with a value judgment placed on it.

In the end, this book asks us to stop and take a look at the world around us from a slightly different perspective—the perspective of measurement. Lord Kelvin looked at his world and declared in no uncertain terms, "To measure is to know." But is there more to it than that? Let's see what we can find out.

ACKNOWLEDGMENTS

Trevor Lipscombe saw something of merit in the early stages of this book, for which I will always be grateful. The patient guidance he provided to a first-time author is much appreciated. Peter Dreyer's incisive editing is likewise gratefully acknowledged. The rest of the staff at the Johns Hopkins University Press has been friendly, professional, and patient.

John Shadley has been a friend and colleague for nearly twenty years. His support helped me finish this book, and he provided useful suggestions on the material on Wallace Sabine in Chapter 2. My department chairman, Ed Rybicki, and dean, Steve Bellovich, granted me sabbatical leave and encouraged me to pursue this project, a relatively unusual one for an engineering academician. Hank Knight's ideas and encouragement helped the book get off the ground. My students at the University of Tulsa deserve thanks as well, for lots of reasons, not the least of which is that they have had to listen, over and over again, to many of the same stories I tell in this book.

My mother and father raised their son in a home where good conversation was the finest pleasure and where words and ideas were things to be cherished—only to watch him grow up and become an engineer. My wife Mia read the manuscript all the way through when it was pretty rough and provided constant support, encouragement, and a whole bunch of good ideas and provocative questions.

Over the years, I have occasionally been told that I "write well for an engineer." This sort of backhanded compliment (not unlike being told that you surf well for a guy from Oklahoma) was, I suppose, among the reasons why I had always wanted to write a book. More of us engineers, especially those who surf better than I do, ought to do the same. If this book inspires any future surfing expeditions by my colleagues in the engineering profession, the effort will have been worth it.

DOES

MEASUREMENT

MEASURE

UP?

CHAPTER 1

OF
LOVE
AND
LUMINESCENCE

What, Why, and How Things Get Measured

> *I often say that when you can measure what you are speaking about, and express it in numbers, you know something about it; but when you cannot measure it, when you cannot express it in numbers, your knowledge is of a meager and unsatisfactory kind.*
> —Lord Kelvin, "Electrical Units of Measurement" (1883)

A Day in the Life

You awaken to the weather forecast on your clock radio. It will be 92°F today, with a heat index of 101. Another scorcher. Oh well, time to get moving. The morning paper says the Dow was up 2% yesterday—good news. Maybe that will help pay college tuition for your daughter. Based on her standardized test scores, she won't get much scholarship help. On the way to work, you glance at the dashboard GPS (global positioning system), just to make sure you're not lost. At the office, you open an e-mail from your mother, who's worried about Dad's blood sugar level. A phone call from the boss interrupts you. Sales and market share are down for the second straight month—not what you wanted to hear. You say a silent prayer that you won't be adding to the unemployment rate. Thirty minutes on the treadmill at lunch push your heart rate near the target maximum. You deserve the pizza-to-go you grab on the way back to the office. How many grams of carbs is that? Let's not go there. On the evening news, more bad news for the governor. His poll numbers are down again, and the opposition has already raised $25 million for the election. Better news for your alma mater's football team: an early win has sent them soaring into the top 10. In the weather, no end in sight for the heat wave. All these triple-digit heat index days are sure to stir up more talk of global warming. You switch off the box and head for the sack. If you go to bed now, you could get almost seven hours of sleep. And tomorrow, as they say, is another day . . .

What Is Measurement, and What Is Knowledge?

We spend a great deal of time measuring, being measured, and thinking about the measurements of others, but I suspect most of us don't often stop to think about the measurement process itself. Just what is measurement? The Czech mathematician Karel Berka distills the measurement process down to three elements. First, there is the object of the measurement. Second, there are the results of the measurement. And third, there are "certain mediating empirical operations." The object to be measured can be literally anything. "All that exists, exists in some amount and can be measured," according to the American psychologist E. L. Thorndike (1874–1949). This can get us into trouble. For example, not everyone believes that intelligence is a thing "that exists," and that it can therefore be measured—although we as a society seek to measure it all the time. But for the moment, let's agree that the measurement process starts with an object, and that for these purposes, intelligence is such a thing.

The second aspect of this measurement process is the results of the measurement. The actual act of measuring (the instruments and procedures required) is not specifically mentioned by Berka—let's just assume it to be a part of this second step. Wrapped up in the "results" of measurement are all kinds of technical details: instrumentation and procedures (as noted above), uncertainty, and other statistical considerations, in short, the kinds of things that frequently turn on people like myself.

Finally, there are those "certain mediating empirical operations." This means just that, frequently, the exact thing that we measure (the result, in Berka's terms) is not directly and immediately useful to us. A magnetic resonance imaging (MRI) scan of your brain is a highly sophisticated measurement. More precisely, it is a great many individual measurements, all presented for your consideration at the same time. The raw, unmediated results of these measurements consist of the absorption and emission of energy in the radio-frequency range of the electromagnetic spectrum by the tissues in your body. This, in a nutshell, is what an MRI measures. It turns out that different kinds of tissue in the human body (both normal and abnormal types of tissue) absorb and emit this type of radiation at different rates. It is hardly surprising that raw measurements of this kind are not something that a human being (even a highly trained physician) can directly make use of. In this case, the "certain mediating empirical operations" required to make the measurements useful are complex indeed and involve a frightening amount of math, which is performed at breakneck speed by a computer. The end result—one of those fancy multicolor images of a slice of your body—is something that a physician can most assuredly make use of. The physical principle behind MRI (specifically, nuclear magnetic resonance) and the technology required to produce the images have resulted in Nobel Prizes in Physics, Chemistry, and Medicine, and in the birth of a multibillion dollar industry. It is a classic "measurement" story and a continuing triumph.

So that is the measurement process: find something to measure (the object), measure it (to produce the result), and then manipulate that result to make it useful (the mediating empirical operations). All this creates knowledge. An MRI machine gives your doctor knowledge about what is going on inside your body. A bathroom scale gives you knowledge about whether your diet is working. A standardized test gives you knowledge about the educational progress of your child. In each case, knowledge comes from measurement.

The Love Meter

It is clear enough that a great deal of knowledge comes from measurement. But can knowledge actually come from anywhere else? Remember back to the first time you fell in love. How did you know it was the real thing? Just where did that knowledge come from? It's a pretty good bet that you weren't measuring anything. And it's also possible that, novice lover that you were, you may have been wrong in your assessment of your true emotions.

Might measurement have helped? If measurements can be used to detect lies, why not love? One could easily construct a survey designed to measure how much romantic love one person has for another. Sample question: "On a scale of one to ten, with ten being 'constantly,' how much do you think about this person?" Simply take the test, tally the results, and find out if you are really in love. "Any score over 90 indicates true love." Such surveys exist, as supermarket checkout line magazine browsers everywhere can attest. Dating agencies also provide surveys to discover your preferences in partners, thus allowing them to rank you and match you with the person of your dreams.

Perhaps measurement tools like the love survey are too pedantic for your tastes, and you'd prefer something a little more organic. We'll paste some electrodes to your skull (and perhaps elsewhere), then, and measure electrical activity as stimulated by thoughts of this person you think that you love, or by their video image, or through some other sensory input. Next, we'll compare your brain activity to similar measurements provoked by a person you don't know, or your boss at work—that is, by someone you definitely don't love. We'll also compare your measurements to those of other people in love.

There may be other ways to measure romantic love as well. For example, Figure 1.1 shows two old-fashioned love meters (also known as "hand boilers") in action.

All kidding aside—this is a serious discussion about measuring love. However, I am not suggesting that it is a good idea to try to measure love, or a fruitful area for further research, or anything of the sort. I am simply suggesting that this is how things seem to go when it comes to knowledge and measurement. Unmeasured qualitative knowledge ("I think I'm in love") is replaced by quantitative knowledge ("I scored 97 on the love survey; I'm definitely in love!"). We will see this over and over again, in

Fig. 1.1. The subject on the right has more love than
the one on the left, according to these "love meters."

sports, business, medicine, environmental issues, and so on, throughout the
chapters of this book. Personally, with respect to love, I hope we never get to
that point. It's much more fun the old-fashioned way!

A Mathematical Definition of Measurement

To a mathematician (in this case, Fred S. Roberts), a measurement is a
"mapping of empirical objects to numerical objects by a homomorphism."

A homomorphism is a "transformation of one set into another that preserves in the second set the operations between the members of the first set." If this makes you glad you are not a mathematician, I suspect you are not alone.

An example comes from the story of Noah in the Bible (Genesis 6). God gave Noah certain instructions for building the Ark ("build it 300 cubits by 50 cubits by 30 cubits"). If and when Noah measured the completed Ark (perhaps for quality control purposes), then, he would have been performing "a mapping of empirical objects" (the size of the Ark) to "numerical ones" (the number of cubits in each of the Ark's dimensions). For this mapping to be "homomorphic," the second set (the number of cubits in each dimension) would have to preserve the fundamental properties inherent in the size of the Ark. Practically, we can think of this as meaning that the dimensions of the Ark in cubits would have to give us a good enough idea of the size of the Ark to enable Noah to do his job. That is, by measuring the Ark in cubits (the distance from the elbow to the tip of the middle finger—and thus not a very accurate unit of measure), Noah ensured himself that he was doing God's will, and that all of the animals would actually fit in the Ark. Evidently, the cubit worked well enough for this, and Noah didn't need a laser-based, computer-controlled measurement system to measure the Ark to the nearest micrometer.

To build the Space Shuttle, on the other hand, or a computer chip, we need a more sophisticated way to measure distances—the cubit won't get the job done. Put another way, the cubit is "homomorphic" (enough) for Noah's Ark, but it is not homomorphic for the Space Shuttle. It does not "preserve the relations in the first set" (those of the Shuttle's designers). The mathematical language remains the same. It is the measurement (the cubit in this case) that is lacking.

For a more abstract example, consider "grades and education." We use grades to measure student progress and mastery of subject matter. What are the "relations in the first set"? We want students to learn, to become productive members of society, and so on. Education is considered crucial in this process. But how do we measure it? Grades and standardized test scores are the measurements of choice. Are they homomorphic? I'll let you decide.

On Measurement and Knowledge

Our lives have become increasingly quantified. Over time, many things that we once considered "unmeasurable" have become routinely quantifiable. Our daily routine is more and more ruled by numbers—by the things we measure. Is there anything that we shall not someday be able to measure, accurately and profitably?

Should you ever sit on a jury, you may be asked to judge someone's guilt "beyond a reasonable doubt." How can you measure that? Might a jury of one's peers someday be replaced by a machine that watches and listens to a trial and then somehow measures the "reasonableness" of guilt? This sounds like science fiction, and if it doesn't make you a little bit nervous, well, maybe it should. On the other hand, some people judged guilty "beyond a reasonable doubt" of horrible crimes are being released from prison these days when DNA evidence proves their innocence. (Conversely, the guilt of a great many others has been confirmed by DNA evidence.) The measured objectivity of DNA evidence presents a stark contrast to the unmeasurable subjectivity of a "reasonable doubt."

As one assesses the state of the art of measurement across the broad spectrum of human endeavors, it is easy to conclude that we are marching inexorably toward a future in which *everything* will be measured.

"Hard" Versus "Soft" Knowledge

"A major difference between a 'well-developed' science such as physics and some of the less 'well-developed' sciences such as psychology or sociology is the degree to which things are measured," the mathematician Fred Roberts observes. Sciences that are well developed in this sense are often called "hard" sciences, while those that are less well developed are referred to as "soft."

Newton was a "hard" scientist in this sense—he measured things. He quantified the motions of the planets and the laws of optics, among other things. Freud was a "soft" scientist. He listened to people describe their dreams, and from that he tried to figure out why they acted in the ways that they did. How soft can you get? If Freud was measuring anything, consciously or unconsciously, it's hard to know what it might be.

In a hard science like physics, measurement is king. Lord Kelvin, after all, was a physicist. When he says that without measurement, all knowledge is "meager and unsatisfactory," you can almost feel the smugness of the hard scientists, as well as the envy of those in the soft sciences. What can *we* measure, many soft scientists throughout history seem to be saying, so that our status in the scientific world will improve? Lots of things, as it turns out.

It is cynical to imply that the growth of measurement in the soft sciences is due only to a desire by those scientists to be taken more seriously. More optimistically, one might say that as the traditionally soft sciences become harder, as they begin to quantify more and more, they are generating more and more knowledge. As Kelvin might say, now they are really beginning to know something, having replaced meager and unsatisfactory knowledge with measurement.

The soft sciences are not as soft as one might have thought, it turns out, nor are the hard ones as hard. Is it possible that there is just one continuum of science, and of knowledge as well? If that is true, then measurement would seem to be intimately related to knowledge in all its myriad forms. In Peter Weir's film *Dead Poets Society* (1989), the students are instructed by their poetry text to determine a poem's "greatness quotient," which is defined as the product of its "perfection" times its "importance." Belittling this attempt at what one might call extreme measurement, the teacher, John Keating (Robin Williams), tells the students, "We're not laying pipe here, we're talking about poetry."

There is perhaps no more eloquent voice than that of Stephen Jay Gould (1941–2002) expressing the idea that no science is really totally objective, whether hard or soft. In his book *The Mismeasure of Man*, Gould criticizes "the myth that science itself is an objective enterprise, done properly only when scientists can shuck the constraints of their culture and view the world as it really is. . . . I believe that science must be understood as a social phenomenon, a gutsy, human enterprise, not the work of robots programmed to collect pure information." Gould argues that "facts" are not objective but are influenced by culture, and that as science changes (advances?) through the years, it does not necessarily "record a closer approach to absolute truth, but the alteration of [the] cultural contexts that influence it so strongly."

On the other hand, Gould reveals that his belief about the nature of science is fundamentally optimistic. "I do not ally myself with . . . the purely relativistic claim that scientific change only reflects the modification of social contexts, that truth is a meaningless notion outside cultural assumptions, and that science can therefore provide no enduring answers." He is a scientist, and thus believes that science, "though often in an obtuse and erratic manner," can discover a "factual reality." Galileo, for example, "was not shown the instruments of torture in an abstract debate about lunar motion. He had threatened the Church's conventional argument for social and doctrinal stability. . . . But the Church soon made its peace with Galileo's cosmology. They had no choice; the Earth really does revolve around the sun."

Higher education, too, is struggling with the relationship between measurement and knowledge, and with the hard-versus-soft question. The president of Harvard, Lawrence Summers, wants to overhaul the undergraduate curriculum at the nation's most prestigious university. Summers wants to do this because he believes the very nature of knowledge has radically changed. As he said in the *New York Times Magazine* on August 24, 2003, "More and more areas of thought have become susceptible to progress, susceptible to the posing of questions, the looking at the world and trying to find answers, the coming to views that represent closer approximations of the truth."

"Progress" and "closer approximations of the truth," in these terms, might be thought of as the ability to measure things in a meaningful way. Summers is essentially saying that we are getting better and better at measuring more and more things, and that this knowledge thus gained, these "closer approximations of the truth" replace our formerly meager and unsatisfactory qualitative knowledge. It follows that the measurements that allow us to do this are of increasing importance in many phases of our personal and professional lives, and that the university student who is unprepared to understand and think critically about these things has not really received much of an education.

Summers describes how modern measurements tools and techniques have become highly refined and ubiquitous. Archaeology, for example, "was at one stage a kind of 'Raiders of the Lost Ark' operation. Now we [Harvard] are hiring a chemist who can figure out diet from fingernail clippings."

In his *New York Times* article about Summers, James Traub frames the debate at Harvard over the changes in undergraduate education that Summers promotes. Great universities, he says, "have traditionally defined themselves as humanistic rather than scientific institutions." Summers believes that, regardless of whether this balance has shifted, or even if it should, that the distinctions between science and humanism have blurred—in favor of the quantitative. "The soft has become harder, rather than the other way around." Faculty members (from Harvard and elsewhere) throughout the hard/soft spectrum seem more worried about these changes than, for example, about changes in the left/right political spectrum on their campus.

It is also possible to conclude that the hard, analytical domains are softer than one might have believed. The important point, however, is that the distinctions between hard and soft, between the humanities and the sciences, *have* blurred. This book argues that it no longer makes much sense to think about these things as completely separate and fundamentally different from one another.

Those of us trained in science and technology are intimately familiar with the relation between measurement and "scientific knowledge." Lord Kelvin's assertion that without measurement our knowledge is meager and unsatisfactory is tacitly understood. Scientists measure things to better understand nature. Engineers analyze, design, and build things to precise quantitative specifications and standards. The "measurement revolution" in these fields is so entrenched that technologists rarely even think about it. Only in its absence does it become apparent, as it did to me as a young professor.

When I first began teaching, I was quite surprised at the lack of appreciation many of my students showed for this—at their level of "nonquantitativeness." I have long since grown accustomed to teaching beginning engineering students always to quantify when, for example, comparing things. Don't tell me that "aluminum is lighter than steel" (or even "a lot lighter"). Tell me instead how dense aluminum and steel are—give me the numbers. That way, I get the relative comparison (which one is lighter) built in right along with the absolute measures, which are far more useful. The comparison (which one is lighter) is helpful up to a point, but I can not only use the numbers, the measurements, to compare the materials; I can put them into equations to actually design something. The nonquantitative relative comparison alone doesn't let me do this. I need the densities of steel and aluminum—and of balsa wood, perhaps—before I can determine how heavy my kite will be and what to build it from to ensure that it will get off the ground.

This measurement revolution has not taken hold everywhere, but it is spreading. As a teacher, I grade my students. In the United States, our custom is to assign letter grades. An "A" is assigned for excellent work, a "B" for good work, and so on. Most teachers (certainly those in science and technology) assign these letter grades based on numerical equivalents. (In many other countries, all grades are numerical, and there are no letter equivalents. They have dispensed with any pretense that grades are non-quantifiable.) Those students with an average score of at least 90% receive an A, 80%–90% a B, and so on. These numbers are assigned based on the results of tests, homework, and reports. If you answer 19 out of 20 test questions correctly, that's a 95%, for example. Even written reports often receive numerical scores. A teacher might deduct, say, 5% for an incorrect conclusion on a report.

But not all teachers think this way about grades. Several years ago I had an engineering student who was enrolled in a philosophy class. Near the end of the term, he was concerned about his grade. On the midterm exam, he had received an A−. On two papers, he had received a B and a B−. Going into the final, he wanted to know how well he would need to perform in order to receive an A for the course. A meeting with the instructor revealed a deep cultural divide related to the culture of measurement. It seemed that there was no answer to the student's question regarding the minimum performance necessary to guarantee an A. "But how many points is an A worth?" the student wanted to know—wishing, like the well-trained engineer that he was, to calculate the grade necessary to achieve a 90% in the course, as he was accustomed to doing in all his science and engineer-ing courses. An A is an A, was the instructor's enigmatic, if not philosophi-cal, answer, and there are no "points" about it.

To that instructor, letter grades were abstract and unquantifiable. It was almost as if someone had asked the instructor to quantify his love for his country, a parent, or a spouse (the love survey notwithstanding). To make an A in my class, the instructor informed the student, "You must do 'excellent work' throughout the semester. If you do excellent work on the final exam, you may possibly receive an A in the course. That will be for me to decide." The student left the meeting more bewildered than ever, and the divide between the "culture of measurement" and that of nonmeasurement grew just a little bit wider.

If you find the above example silly, you are probably not alone, especially if you are young. Letter grades in school are much more quantitative than they once were. Other examples of this lack of quantification are discussed in more detail elsewhere in this book.

The Darker Side of Measurement

There is a darker side to measurement, which rears its ugly head from time to time. It manifests itself in a variety of ways, perhaps nowhere more insidiously than in "measurements of the mind" (Chapter 6). If intelligence is an object, then, in Thorndike's terms, it ought to be measurable. If we can assign a single number to the intelligence of each person, we can rank those numbers in order from highest to lowest. From this, all kinds of mischief may arise, particularly if it is further concluded that the thing called "intelligence" is something we are born with, and that we cannot change. Indeed, all kinds of mischief *have* arisen from this particular example. The story of the measurement of intelligence is only one example of how measurement can be misused. Others will be recounted in turn in various other chapters.

The darker side of measurement might be divided into two aspects. The first is the misuse or abuse of measurement, as exemplified in the case of intelligence. The second aspect is simply the overuse of measurement—the indiscriminate application of measurement, the overquantification of the world, or whatever you want to call it. Many scientists, particularly those from the "hard" sciences, have remarked on this over the years.

There exists in American society a certain tendency toward excess. If a little bit of something is good, then, we feel, a lot is certain to be better. It's that way with the sizes of cars, houses, and restaurant meal portions, and thus it seems to be with measurement. The ability to measure has resulted in innumerable improvements in many important and disparate areas of our society, including business, medicine, sports, criminal investigation, and many others that are described throughout this book. If attempts to impose measurement on other areas, such as intelligence, have proven less fruitful, and even damaging, is it necessarily the fault of those who attempt to do the measuring, or should the blame lie with those of us who *apply* those measures, blindly, in inappropriate ways?

The British immunologist Sir Peter Medawar (1915–87) writes of the differences between natural sciences and what he calls "unnatural" science. The chief difference is that unnatural scientists believe that "measurement and numeration are intrinsically praiseworthy activities (the worship, indeed, of what the art historian Ernst Gombrich called *idola quantitatis*)." This would at first glance appear to be incompatible with Kelvin's belief that "to measure is to know." However, I do not believe that Kelvin would have found the indiscriminate application of "measurement and numeration" to be "intrinsically praiseworthy," any more than a good surgeon believes that every patient needs an operation, or a scrupulous lawyer that all his clients should file a lawsuit against someone. The indiscriminate use of measurement for its own sake would have been anathema to Kelvin.

In spite of the growing pervasiveness of the measurement culture and the clear dangers of the darker side of measurement, I remain optimistic about the future of measurement and its place in our society. For one thing, I believe it is inevitable that the measurement culture is going to gain an increasingly tight grip on more and more aspects of our society. In the face of that, it's nice to believe that these changes will be mostly for the good. On the other hand, I also believe that it is important that we as a society be aware of the extent to which the measurement revolution controls our lives. We need to know, too, what these measures mean, where they come from, how they are applied, when to believe them, and when to ignore them. As an educator, I have to believe it is better to light a single candle (all the while measuring its luminous intensity) than to curse the darkness (which we can also quantify).

CHAPTER 2

DOING

THE

MATH

Scales, Standards, and

Some Beautiful Measurements

Count what is countable, measure what is measurable,
and what is not measurable, make measurable.
—attributed to Galileo Galilei (1579)

Mathematics and Measurement

Mathematics is not measurement per se. But the two are certainly related, and it is even possible that the birth of mathematics as a formal science grew out of a desire to make more precise measurements. In ancient Egypt, land surveyors may have begun the formal study of geometry in an attempt to improve the precision of their work.

Writing in *Science* in 1958, the psychologist S. S. Stevens said, "Given the deeply human need to quantify, could mathematics really have begun elsewhere than in measurement?" Like geometry, arithmetic was probably also invented for purposes of counting and measurement. Our practical need to count our possessions, particularly when it came to the exchange of goods and services, led directly to arithmetic. Clay tablet accounting dates to at least 3000 BC.

Mathematics greatly expands the utility of measurement, as the early geometricians and arithmeticians surely discovered. For example, if I measure the diameter of a circle, I can easily calculate the area of that circle using the simple geometric relationship $A = \pi r^2$, where A is the area and r is the radius or one-half the diameter. Thus, I have measured the circle's area—indirectly, perhaps, but I have measured it nonetheless. I once had a professor who chastised his students when they referred to this as a "measurement." It's a calculation, he would say. You can't measure the area of a circle this way, you can only calculate it from its diameter. Technically, he was right, but in my opinion, this is a bit pedantic. The area of the circle is based entirely on a single measurement—that of the diameter—the value obtained for the area is therefore only as good as the measurement of the diameter. To say that we have thus measured the area of the circle is perhaps to oversimplify things a bit. But this is an important example. It turns out that most of the measurements we come across in our lives these days are not pure measurements. They are rather the results of some kind of manipulation of one or more pure measurements.

We shall look at lots of examples of measurements where the results that we see (often on a computer screen) are sophisticated mathematical manipulations of measurements that are, in and of themselves, of limited utility. A digital photograph is a particularly graphic (no pun intended) example. In addition, the branch of mathematics known as calculus is almost infinitely useful in enhancing and extending the utility of measurements. For example, if I measure the velocity of something versus time, even if that velocity changes wildly, I can obtain the total distance traveled by the object using a simple application of calculus—in this case an integration. Computers are great at integrating things and at performing all sorts of other mathematical manipulations of measurements that are designed to make them more useful to us all. Combine a computer with the ability to do calculus and other kinds of sophisticated math and you gain the ability to make measurements really useful.

Measurement Scales

If you take a college course on measurement theory, or perhaps a course on statistics, you are likely to be informed that any measurement can be classified as one of the following four types: nominal, ordinal, interval, or ratio. This classification was first proposed by S. S. Stevens in 1946. The differences among the scales are important in part because the type of scale determines to some extent what you can and cannot do with a certain measurement. This will probably make more sense as we discuss the four scales individually below.

Nominal data are classifications for which it doesn't make sense to order things. The classic example is classifying (the subjects of an opinion poll, for example) as "male" or "female." We can count up the number of each and calculate, say, the percentage of males in an opinion poll who are in favor of reelecting the president. But that's about all we can do, mathematically, with male/female data. It makes no sense, for example, to say "male is greater than female." Male is *different* from female, but that's as far as we can go with nominal data.

Ordinal data are ordered, so that "greater than" and "less than" make sense, but the differences between values are really not important. Restaurant ratings (such as those in the *Michelin Guide*) are a good example. We can say that a two-star restaurant is better than a one-star restaurant, but we can't really quantify the difference between the two using this ordinal measurement scale.

Interval data are ordered and lie on a scale such that differences are quantifiable. The Fahrenheit temperature scale is a good example. I can observe that 80°F is 30°F hotter than 50°F, just as 210°F is 30°F hotter than 180°F. That result makes sense. But I cannot divide 100°F by 50°F and draw the conclusion that 100°F is "twice as hot" as 50°F. This is because the Fahrenheit measurement scale does not have a "natural zero." Zero degrees on the Fahrenheit scale is an arbitrary point (just as it is on the Celsius scale), and ratios like this therefore do not make sense.

Ratio data, as the name suggests, are like interval data, except that ratios do make sense. The Kelvin (°K) temperature scale (the "absolute temperature" scale) is a ratio scale. It has a natural zero because 0°K is the coldest possible temperature, absolute zero. Other ratio data scales are things like height and age. Someone who is forty years old is both twenty years older (interval) and twice as old (ratio) as someone who is only twenty.

Stevens did more than just classify all measurement into these four categories. He also identified the different types of statistical analyses that can and cannot be performed on each type of data. Some statisticians and others have a variety of disagreements with Stevens's particular statistical conclusions, but that needn't concern us here. The four basic categories, nominal, ordinal, interval, and ratio, remain a useful way to categorize measurement, and part of being "numerate" (the numerical analogue to being literate), I believe, is having some facility with these measurement scales.

At the risk of muddying the waters, lots of important things are measured on a logarithmic scale, which at first glance doesn't seem to fit any of the above categories. Three great examples of logarithmic scales are the decibel scale for loudness, the Richter scale for earthquake intensity, and the pH scale for the strength of acids and bases—but there are others as well. It turns out that a logarithmic measurement scale is really just a ratio scale, except that it obeys the mathematics of logarithms, rather than simple arithmetic. One way to understand a logarithmic scale is to turn things around—that is, to make a logarithmic scale out a common everyday linear scale.

Let's use length as our example of a linear scale. On a meter stick (the metric equivalent of a yardstick), there are 100 centimeters marked off at equal intervals. Each centimeter is further subdivided into 10 millimeters. If I were to create a logarithmic meter stick, I would first label equally spaced points 1, 2, 3, 4, and so on, just like a regular, linear meter stick. The difference is, on the logarithmic meter stick, the point labeled "1" would correspond to 1 centimeter, "2" would correspond to 10 centimeters, "3" to 100 centimeters (1 meter), "4" to 10 meters (1,000 centimeters), and so on. Thus, on the logarithmic meter stick, the difference between 1 and 2 is "ten times," the difference between 2 and 3 is "ten times," and so on.

Representing everyday lengths on a logarithmic meter stick would be a pain, but you could do it. For example, if I were to represent my own height on a logarithmic meter stick, I would come up with a value of about 2.28. So how tall am I? Well, if I raise the number 10 to the 2.28 power (grab a calculator) that equals about 190.5 centimeters (6 feet 3 inches), which is how tall I am.

We're obviously much better off representing people's heights on a standard linear scale, instead of a logarithmic one. This is because the difference in height from one person to the next is never all that great. I could measure everyone on the planet from babies to giants with one stick that's, say, 2.5 meters (9 feet 2 inches) long.

But what if I wanted to represent the lengths of all kinds of other things with a single stick? How could I represent the height of an ant (1 millimeter, or 0.1 centimeters), a person (180 centimeters), a Giant Sequoia tree (nearly 100 meters, or 10,000 centimeters), and the distance from the Earth to the Sun (about 150,000,000 kilometers, or 150×10^{11} centimeters) on a single scale? If I try to lay all these measurements out on a linear scale (a really long one!), the ant, the person, and the tree will all essentially lie at "zero" on the scale. Even though the tree is 100,000 times as tall as the ant, in terms of this astronomically long linear meter stick, the ant and the tree are both "zero."

On the other hand, if I put the ant, person, tree, and Earth-Sun distance on a logarithmic scale, the ant lies at minus 1, the person at 2.25, the tree at 4, and the Earth-Sun distance at 13.18. They all fit nicely on one "stick" and I can easily distinguish among them all (Fig. 2.1).

This is pretty much how it works for things like loudness and earthquake intensity. They are much more logically represented on logarithmic scales because the differences in the quantities we measure are so huge. The Richter scale for earthquake intensity uses a logarithmic scale because there is such an enormous variation in the seismic energy of earthquakes. Because really big earthquakes contain so much more energy than smaller ones, we could never distinguish among the smaller ones if we used a linear scale (just as we could never distinguish the size of an ant from that of a giant tree on an astronomical linear distance scale). So a magnitude six earthquake is ten times more energetic than one of magnitude five on the logarithmic Richter scale. (As a brief aside, the magnitude of an earthquake, on the Richter scale, is *not* a measure of how destructive a quake is. Generally speaking, the greater the quake's energy, as measured on the Richter scale, the more damage it will do, but this is not always the case.)

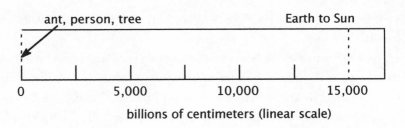

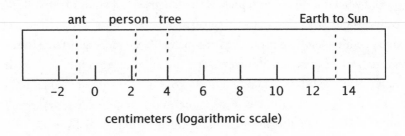

Fig. 2.1. Length measurement using linear and logarithmic scales

A variety of things related to our senses (such as loudness) are best measured on logarithmic scales as well. Having observed that the internal measuring systems we were born with (our "sensations") do not operate on the same scales as those of the "stimuli" measured on our human-made instruments, the German psychologist Gustav Fechner formulated what is called Fechner's Law (1860), that "as stimuli are increased by multiplication, sensation increases by addition." For example, using acoustic instruments, we can measure the stimulus of a sound in terms of the amplitude of the pressure wave the sound creates. The sensation (of loudness, in this case) created in a human being by that stimulus can only be measured somewhat subjectively, by questioning that person. It turns out that doubling the stimulus of a sound does not double the sensation of loudness. As the sound stimulus is multiplied (doubled, quadrupled, etc.), we only add incrementally to the sensation of loudness. Put another way, as our sensation of loudness (what we perceive from ear to brain) goes up in equal steps, the sound stimulus necessary to produce that sensation goes up in multiplicative steps. Thus the familiar decibel (dB) scale of loudness is a logarithmic scale. A sound that measures 70 dB is perceived as twice as loud as a sound of 60 dB.

Fechner was a pioneer in the field of psychophysics, which attempts to measure subjective aspects of the mind, just as we measure other physical phenomena. Things like the loudness of a sound, the brightness of a light, the intensity of a taste (sweetness or saltiness, for example), and various tactile sensations (such as the response to electric shock) are nonlinear. That is, just as with the sound example above, if I make a graph of the stimulus versus the sensation, I do not get a straight line. Some of these phenomena (like loudness) are best represented on a logarithmic scale, while for others, the mathematical form is a little different (as noted by Stevens and others).

In terms of measurement scales, we simply try to measure things in ways that make the most sense. That is, we measure them in ways that give us the most knowledge about whatever it is that we are measuring. Regardless of the scale that we are using, the "factual reality" that exists is discovered or confirmed through measurement. Let's look at a few examples.

The Speed of Light

I find certain measurements absolutely beautiful—in the same way that I might find a painting or a musical performance beautiful. The story of the measurement of the speed of light is a good example of the beauty (combined with utility) that can be found in scientific measurement. Early philosophers such as Aristotle and later Descartes believed that light traveled instantaneously. Galileo knew that light travels much faster than sound (he saw the flash of a distant cannon well before hearing its boom), but he was smart enough to know that this did not mean that light moved with infinite speed. Galileo suggested an experiment to measure the speed of light that was ingenious but unfortunately also impractical. At night, two people would stand on distant hilltops, holding covered lanterns. One person would uncover his lantern, and then as soon as the light was visible to the second person, the latter would uncover his lantern. By measuring the distance between the two lanterns, and the difference in time between the successive uncoverings of the lanterns, the speed of light could be determined. A similar technique can be used to measure the speed of sound by timing an echo.

Measuring the speed of light this way is a fine idea, but no one is that quick! Even today it would be difficult to measure the time lag, which would be a tiny fraction of a second even for hilltops miles apart. We can think of Galileo's idea as a thought experiment, rather than a practical one. Even though Galileo claims to have tried his experiment at distances less than one mile (at those distances he was unable to measure a time lag), it's quite likely that he knew that this was not a practical experiment. But it did what a good thought experiment should do, it got people thinking!

The first practical measures of the speed of light, about 50 years after Galileo, were astronomical. When successive eclipses of one of Jupiter's moons came either earlier or later than expected, the Danish astronomer Øle Roemer correctly concluded in 1676 that this was because the distance between Earth and Jupiter was changing, and that light was thus taking different amounts of time to travel from Jupiter to Earth. His rough estimates of the speed of light from these measurements were refined by Isaac Newton in his *Principia*, who concluded that it takes light seven or eight minutes to travel from the Sun to the Earth—which is a good estimate.

Others used various astronomical observations to further refine measurements of the speed of light, but it wasn't until about 1850 that the search for the speed of light came back to Earth, and back to Galileo's experiment with the lanterns. Two Frenchmen, Armand Fizeau and Jean Foucault, independently devised and carried out similar experiments that combined the spirit of Galileo's lanterns with ingenious (and beautiful) techniques for dealing with the extremely small time lags involved. (When Galileo placed his lanterns one mile apart, he was essentially trying to measure a length of time less than one one-hundred-thousandth of a second. Not even Muhammad Ali in his prime was that quick.) But both Fizeau and Foucault found mechanical means to get around this problem. Foucault built a rapidly rotating mirror onto which he focused a beam of light (the mirror was spinning about an axis in the plane of the mirror, the way a paddlewheel rotates). At one precise point during each revolution, that mirror reflected the light directly at another, stationary, mirror on a distant hilltop. The second mirror reflected the light back on the first mirror, which by then had rotated ever so slightly. This reflected light thus bounced off the first mirror at a slight angle. The angle between the outgoing and incoming beams of light, however, was big enough to be measured accurately, and from this the speed of light, accurate to within 1%, was calculated. Voilà!

In 1880, a young instructor at the United States Naval Academy, Albert A. Michelson, wanted to include more demonstrations in his lectures. It seemed as though it would be interesting to reproduce the experiments of Fizeau and Foucault, and Michelson persuaded himself that he would not only be able to gin up a really interesting demonstration but also greatly improve upon the respectable accuracy the Frenchmen had achieved. Bankrolled to the tune of $2,000, a tidy sum in those days, by his wealthy father-in-law, Michelson set out to have the last word on the speed of light. Using extremely refined mirrors and lenses, and measuring the 2,000-foot distance between mirrors to within one-tenth of an inch, Michelson was able to measure the speed of light to within 30 miles per second of the value we accept today (which is about 186,000 miles per second). This was more than 20 times more accurate than Foucault, and it helped make Michelson a famous man. In 1884, Michelson took part in a series of famous seminars given by Lord Kelvin at the Johns Hopkins University. Kelvin breakfasted each day with Michelson, Edward Morley, Lord Rayleigh, and others, and then frequently gave spontaneous lectures in the afternoon on subjects arising from their breakfast discussions. In 1907, Michelson became the first American to win the Nobel Prize, for a different but related set of experiments. His measurement of the speed of light was not improved upon for another 40 years.

The experiments carried out to measure the speed of light had utility far beyond the measurement of this fundamental physical constant, however. These experiments were part of the catalyst that got a young German-Swiss physicist named Albert Einstein thinking about a subject we call relativity. A beautiful measurement may well inspire a beautiful theory.

Room Acoustics

Devotees of classical music love their music halls. The same music, even played by the same musicians on the same instruments, sounds different depending on the hall in which it is being played. Why is this? Perhaps more important, how do we know how to design a performance hall so that it "sounds good" no matter what the type of performance, from the spoken word to opera, to musical theater, to various kinds of instrumental music? How do we measure "sounds good"?

The advent of modern-day measurement in the field of the acoustics in public halls is generally acknowledged to have been brought about by Wallace Clement Sabine, a professor of physics at Harvard. In 1895, Sabine was approached by the president of Harvard, Charles W. Eliot, with a vexing problem: a new auditorium on the Harvard campus, the Fogg Lecture Hall, had recently been constructed at considerable expense. However, early events in the hall, including lectures and musical performances, had been less than satisfactory—numerous patrons had complained of the "poor acoustics." Eliot gave Sabine the task of investigating *why* the hall sounded the way it did, and of determining what, if anything, could be done about it.

Sabine approached the problem scientifically. That is, he ran controlled experiments. The instruments with which he made his measurements consisted only of his ears, a pocket watch, the pipe organ installed in the hall, and a team of assistants, who helped him by rearranging various aspects of the hall per his instructions.

Sabine wanted to understand, quantitatively, how the acoustics of the hall changed when his assistants removed all the soft fabric cushions from all the wooden seats in the hall, for example. With the hall in a given configuration (say with the cushions in place) he would play a short note on the 512 Hertz (Hz) organ pipe (about high C) and then, using his pocket watch, record the time it took until he could no longer hear the sound. Today, that length of time is known as the "reverberation time." Then he would repeat the experiment with the cushions removed. He found that the way the hall was configured (cushions / no cushions, windows or curtains open or closed, etc.) significantly affected the reverberation time. Using only these rudimentary instruments (although his ear must have been finely tuned indeed), Sabine gathered voluminous data from the hall. The constants in the equations he established were admirably precise—modern instrumentation has only changed them by about 1%. Based on his work, Sabine was able to make recommendations to Harvard's president as to how the hall should be configured for various types of performances in order to optimize the sound for each occasion.

The spoken word, opera, chamber music, orchestral music, and so on, each require a hall of different acoustic properties to be heard to best advantage. The reverberation time of the hall is one way to measure these properties. A reverberation time of about 1 second is best for the spoken word (a speech or lecture), for example. Less than that, and the voice appears flat and colorless. More than that, and the speaker's words begin to run together and are difficult to distinguish. Longer reverberation times are required for instrumental music (over two seconds for a large orchestra hall). Opera, with its blend of voice and music, lies somewhere in between.

In this way, Sabine for the first time established a system for measuring the acoustic properties of an auditorium, enabling the designers of modern halls to create acoustic environments adaptable to a variety of uses. Since a performance hall is so expensive to design and construct, modern halls must be multipurpose. Through the use of curtains and other removable fixtures, the reverberation time can be changed.

Sabine's quantification makes possible knowledge—about the sound of a performance hall—of a different kind than had existed before. Before Sabine, people knew when a hall sounded good or did not, but they were unable to quantify their knowledge, and they thus found it difficult to improve the situation. Their knowledge was "meager and unsatisfactory": if they wanted to build a new hall, they could either try something and see if it sounded good or design something just like that other place that sounded good. Sabine changed this. In place of the qualitative ("that sounds good"), he substituted measurable quantities, such as reverberation time. Furthermore, by quantifying the sound absorptance of various materials, Sabine gave us a way of predicting the effects of changes on the sound of a performance hall. This allows us to evaluate various options: "What if I put up curtains at the back of the hall? Will it sound good during an opera?" Sabine helps us answer this question. We know the optimum reverberation time for opera—generally somewhat shorter than for instrumental music, for example—and we also know the effects of curtains on reverberation time. Thus, we can predict the effect of curtains on the acoustics of an opera house. Curtains at the back of the hall shorten reverberation time—and this can be quantified. Meager, unsatisfactory qualitative knowledge has been replaced by useful, practical quantitative knowledge. This is a fundamental property of good measurement.

That is not to say that Sabine and his successors in the field of acoustics made it easy to design a performance hall today. The task remains a difficult one, a fact that is borne out whenever an expensive, celebrated new auditorium, such as Barbican Hall in London, opened in 1982, turns out to have poor acoustics.

Fracture Mechanics

The United States and its allies won World War II for many reasons. Among these was the advantage in natural resources (people and matériel) that we are blessed with in the United States. But how to get these resources to Europe, Africa, and the various locations in the Pacific Ocean where the war was being fought? We needed a lot of ships, and we needed them fast. Enter the Liberty ship, a vessel designed to be constructed in an astonishingly short period of time—just a few months. We needed to construct these ships faster than they could be sunk by German U-boats. By sheer weight of numbers, we would thus overwhelm the enemy and get our soldiers, weapons, and supplies where they needed to be to fight the battles and win the war.

The Liberty ships, overall, were a great success. But they did have one perplexing problem: a small percentage of them had the annoying habit of breaking in half, seemingly for no apparent reason. (Nineteen such catastrophic failures were recorded, along with more than a thousand less serious incidences of brittle fracture, out of over 2,700 Liberty ships constructed between 1941 and 1945.) At least one ship even broke in half while it was sitting quietly in harbor in calm weather. Unraveling this mystery was one of the factors that eventually led to the widespread practical application of a branch of science and engineering called fracture mechanics, whose birth can be traced back to the early 1900s.

Engineers design structures to withstand various levels of stress and strain over long periods of time. This is a highly quantitative undertaking. Engineers are further charged with ensuring the performance of their structures under various worst-case scenarios. One such example is the presence of a growing crack or flaw in a structure. Until the 1940s, engineers had very little ability to quantify the performance (safety) of their structures against the presence of cracks or flaws.

A critical aspect of designing anything is putting numbers on—quantifying—the performance required from the finished article. Think of the chair you are now sitting on. If too much weight is placed on the chair, it will collapse. If you tell the engineer a chair must support 300 pounds, she will design it one way. If it must support 1,000 pounds, the design will be different, perhaps radically so. Engineers design things to quantitative specifications for many reasons—not just the weight they can carry. They can design things to quantitative limits on noise, environmental emissions, cost, color, surface finish, and efficiency, to name just a few. Prior to the advent of fracture mechanics, however, engineers could not quantitatively design structural components to operate safely when they contained cracks or other flaws.

Classical solid mechanics allows the engineer to design components (such as a chair) to carry certain loads safely. Fracture mechanics allows the engineer to quantify the effects of cracks or other flaws that may exist in those same structures. What this means in practical terms is that an engineer now has the tools to say with confidence, for example, "that airplane component is safe as long as it doesn't contain any cracks longer than one-half inch."

"As brittle as glass" is an expression with which most of us are familiar. If you have ever worked with glass, you are probably aware that you can break a piece uniformly by making a shallow scratch on it. This is called "scoring." First you score the glass, then you bend it, and (if you scored it properly), it breaks uniformly right along the score line. Working with glass in the 1920s, A. A. Griffith was essentially able to quantify this phenomenon. That is, he was able to measure just how much load a piece of glass could withstand without breaking as a function of the depth of a scratch on the glass. In short, he was able to quantify, "as brittle as glass."

The work of Griffith was the beginning of fracture mechanics. Griffith's theories were gradually tested, refined, and expanded, until today, many components (including critical parts of airplanes) are designed to be able to withstand cracks up to a certain size safely.

The Liberty ships came along well after Griffith had finished playing with broken glass. From a fracture mechanics perspective, the steel from which these ships were made was relatively brittle, especially compared to modern steels. Although not nearly as brittle as glass, the steel in these ships was susceptible to fractures induced by small cracks in the steel, just as a piece of glass fractures along a score line. This had been a problem for steel ships all along. The most famous ship-fracturing incident of all was the sinking of the S.S. *Titanic* in 1912. Brittle steel is generally agreed to have been at least partly to blame for the rapidity with which that famous vessel broke apart and sank after hitting the iceberg.

But a number of Liberty ships broke in half under far less stressful conditions than violent impact with an iceberg. One technological advance in the Liberty ships that allowed their rapid construction was the liberal use of welding, instead of riveting, to join the plates of their hulls. This turned out to be part of the explanation for the breaking-in-half problem. Before the all-welded Liberty ships, the hull plates of steel ships riveted together. Even if they had been constructed of brittle steel, riveted construction gave these ships a built-in resistance to the breaking-in-half phenomenon. If a hull plate of a riveted ship contains a large crack, and something stresses the plate (a big wave hitting the ship, perhaps), then that one plate might break in half. But only one plate breaks—not the whole ship. In a riveted ship, it is very unlikely that a fast-growing crack will jump from one plate to the next—the crack simply stops at the edge of the plate. Because of the violence of its collision with the iceberg, the riveted hull of the *Titanic* was an exception. But the hull of a welded vessel such as a Liberty ship is essentially one giant plate. Fast-growing cracks jump easily from one plate to the next—right across the welds. This was a new phenomenon, unanticipated by the designers of these welded ships.

Modern steel ships feature all-welded construction. Why don't they break in half? Well, they certainly can (and have), but, for one thing, modern steels are far less brittle than those used on the Liberty ships. Today's steels are as a result more able to withstand rough seas and other difficult conditions, even when they do contain cracks or other flaws. This is part of the legacy of the Liberty ships.

Before fracture mechanics, an engineer's knowledge of how to deal with a cracked component was meager and unsatisfactory. He could play it safe and replace the component or leave it alone and hope for the best. The quantification of fracture mechanics allows engineers to evaluate the safety of a crack of a given size in a given component. We really "know something" about it. Fracture mechanics isn't an exact science, whatever that might be. That's okay—engineers can deal with the uncertainties inherent in a fracture mechanics analysis. The point is, we know more with fracture mechanics than we did without it; hence, we are better equipped to understand and deal with the phenomenon of cracks in components.

Measurement in the Social Sciences

In 1869, future President James A. Garfield, while he was still a congressman, made the following observations to the House of Representatives:

> The developments of statistics are causing history to be rewritten. Till recently, the historian studied nations in the aggregate, and gave us only the story of princes, dynasties, sieges, and battles.
>
> Now, statistical inquiry leads him into the hovels, homes, workshops, mines, fields, prisons, hospitals and all other places where human nature displays its weakness and its strength. In these explorations he discovers the seeds of national growth and decay, and thus becomes the prophet of his generation.

Garfield, who developed a proof of the Pythagorean theorem still used in many geometry textbooks, was prescient. During the twentieth century, Americans began to measure almost everything about their social lives. These included things that had been measured before, such as the population, the size of the army, and so forth, as well as a whole host of things that had never been systematically measured, in such disparate areas as crime, love, food, and religion. For example, it was measured that over the course of the century, the percentage of female lawyers in America rose from 1% to 29%, and that of female engineers from less than 0.1% to 11%. Meanwhile, too, the percentage of the American population living in households of six or more fell from 50% to 10%, and the average work week for factory workers declined from 53 to 42 hours.

Not only did we begin to measure these things, but, more and more, we began to talk about them. Presidential candidates spouted complex statistics during debates. Measurements such as the gross national product, the Consumer Price Index, the unemployment rate, and the teen pregnancy rate were all popularized, if not invented outright, in America during the twentieth century. These measurements and others like them became the stuff of everyday discourse, from the concrete canyons of Wall Street to the mom-and-pop diner on Main Street.

These types of social measurements do not yield certitude any more or less than the more scientific measurements discussed earlier in this chapter do. Those who do the measuring and the interpretation of the measurements often disagree about what the numbers mean—as anyone who has ever watched a presidential debate could tell you. In terms of the social sciences, however, "on balance, the explosion of numerical investigation has indeed offered great value: an imperfect measure of accountability, an imperfect way of problem-solving, and an imperfect way of seeing that many things we thought were so, weren't."

That last sentence, a fine, optimistic (and realistic) way of looking at measurement in general comes from *The First Measured Century*, a compendium of twentieth-century American statistics. It doesn't really matter that this sentence is contained in the preface of a book filled with data mined from the social sciences. Lord Kelvin himself, I believe, would be proud of the sentiments therein. Measurement is never really perfect. Any scientist or engineer will tell you that it is impossible to measure anything perfectly. I can never know exactly how much I weigh, or exactly how tall I am—nor can I ever know exactly what percentage of teenaged American girls smoked cigarettes in 1995. But I can easily obtain measurements for those things that are more than accurate enough for nearly any purpose.

One way to gain an appreciation of these changes—of the increasing importance of measurement in society—is to compare old newspapers to modern ones. Newspapers have changed in many ways, and not all of them are related to measurement. Modern newspapers are filled with photographs, for one thing. News articles in older newspapers also tended to be much longer than they are today.

With respect to measurement, one of the big changes in modern newspapers is the prevalence of graphs. The *USA Today* newspaper has gotten a lot of the credit for this, but if it hadn't pioneered the practice, someone else would have. In the old days, graphs weren't printed for the same reason photographs weren't—it was too difficult and expensive. As typesetting and printing moved into the modern age, photos became popular in newspapers well before graphs. Eventually, as predicted by Garfield, the results of "the prophets of their generation," those who did the measuring, found their way into the newspapers. The way those measurements are presented is important. A bar chart comparing the average salaries of major league players in basketball, football, baseball, and hockey, for example, communicates the differences in those figures much more clearly than simply presenting the numbers.

A sample edition of the *New York Times* features dozens and dozens of graphs. Along with predictable inclusions such as graphs of the Dow Jones Industrial Average, we find, for example, bar charts showing the number of players traded per year in the National Basketball Association, pie charts portraying the spending habits of the rich, and graphs delineating the "market share" of all on-line searches by Google, Yahoo, and others.

If you dig up a copy of the *Times* from 100 or 150 years ago, you will not only find no graphs, but little evidence of any kind of the modern measurement revolution. Today, newspapers play a less important role in our society than they did before the advent of radio, television, and the Internet. But newspapers continue to reflect our culture. The news, along with much of the rest of our society, is dominated more and more by measurement.

The Standardization of Measurement

Throughout the history of measurement, standardization has been a recurring theme. The cubit (our word comes from the Latin *cubitum*, elbow, or forearm) was first standardized in about 2190 BC. Traditionally, before it was standardized, one cubit was the distance from the elbow to the fingertips. But whose elbow and fingertips? That was the problem. The cubit was a practical measure for folks like carpenters and merchants. You could measure lots of practical-sized things with it (such as arks), and it was difficult to misplace your measuring stick, attached as it was to your shoulder. However, while my cubit is about 21 inches (I measured it), yours is likely to be different. Sometimes, back when the cubit was an important measurement unit, those differences were insignificant, but sometimes they weren't. Imagine a really short-armed cloth merchant, for example. He could measure out and sell you ten cubits of fabric that might only be nine cubits if you did the measuring yourself.

A standardized cubit was needed, and in about 2190 BC, one was forthcoming. There are actually four standardized cubits still on the books today. Officially, one Egyptian cubit is about 17.7 U.S. inches. An ark built using

Egyptian cubits would have about 5,333 cubic yards in which to store all those animals. An ark built using my 21-inch cubit would have a roomy 8,900 cubic yards for the giraffes to stretch out in.

Money and the Decimal System

The United States of America was among the first countries to standardize and adopt a decimal system of measurement, in 1785. That system, promoted by Thomas Jefferson when he was a congressman and implemented through the efforts of Alexander Hamilton and others, was not designed to be applied to our system of measuring distances or weights. For that, we continue to use the distinctly un-decimal system known as the "U.S. Customary System." The successful decimal system adopted by the United States in 1785, and still in use today, was our system of currency. When it comes to money, we Americans are nothing if not practical.

Imagine what things would be like in the United States if Congress had refused to adopt Jefferson's decimal system of currency, and if we instead used a system similar to our measurement system for, say, distance. When we measure distance in the United States, there are 12 inches in a foot, 3 feet in a yard, and 1,760 yards in a mile. When we need to subdivide the inch, as on a yardstick, we use binary (½, ¼, ⅛, etc.) instead of decimal (¹⁄₁₀, ¹⁄₁₀₀) divisions. Let's invent the following analogous system of currency: In our made-up system, we have four denominations of paper money: the smallest bill is worth $1. There are also bills worth $12, $36, and $1,760. For everyday transactions, we also need coins. Let's have five of them, worth $½, $¼, $⅛, $¹⁄₁₆, and $¹⁄₃₂. Now, let's say you go into the grocery store and when you check out, your total is $67.72. You reach into your wallet to pay, and . . . What combination of bills and coins do you need? If it were me, I'd forget the cash and just put it all on my credit card! (If you did want to pay with cash, you could whip out one $36, two $12, and seven $1 bills, along with one each of the following coins: $½, $⅛, $¹⁄₁₆, $¹⁄₃₂.)

Readers of a certain age will recall that England had a system of currency nearly this incomprehensible until 1970, when it finally adopted a decimal system. Tourists visiting England in the pre-decimal era were frequently baffled by the system (just as foreigners visiting the United States today are frequently baffled by feet, gallons, and the like). When paying for purchases, these tourists would often just place a handful of British money on the counter and trust the clerk to count out the correct amount.

It is interesting that the United States was right out front adopting a decimal system of currency well before other nations. The French even used Jefferson's currency system as a model when developing the metric system, which was first used around 1800. But the United States never followed up with a similar reform of its other weights and measures: we still have inches, feet, ounces, pounds, tablespoons, cups, and gallons. Other than annoying arrantly pedantic professors like me, what difference does all this make? Well, consider the following.

One of my favorite little in-class exercises is to ask a class full of students to estimate the volume of their bodies. I never give any hints when I do this, I just say something like, "We all know how tall we are, and how much we weigh, but how many of us know what our volume is? Your job is to estimate the volume of your body using just pencil and paper, and a calculator if you need it."

High school students in the United States frequently have no idea how to begin this problem (although many do just fine). Undergraduate engineering students (my usual audience) usually approach the problem in one of two ways. The first approach is to assume their body is a perfect cylinder with a height equal to their actual height and a diameter that they are forced to guesstimate, since, of course, their body isn't really a perfect cylinder. (This sort of simplifying assumption often seems strange or comical to those who aren't technically trained, but it is common in all fields of engineering.) You can get a pretty good answer with the cylinder approximation—if you do a good job of guesstimating the diameter of the hypothetical cylinder! The volume of a cylinder is just its height times π times the radius of the cylinder squared. If my height is 6 feet 3 inches and I guess the diameter of my "cylinder" to be 10 inches, then my estimate of my volume is 6.23 feet times π times $(0.417 \text{ feet})^2$. If I plug this into my trusty calculator and do a few units conversions, this tells me that my volume equals 25.4 U.S. gallons.

The second common approach to this problem (and the better one in my opinion) is to use the definition of density (density equals mass divided by volume). If one makes (once again) a simplifying assumption that one's body has the same density as water (not bad, since our bodies are mostly water, and since some of us float in water, and others sink, we must on average be close to the density of water), one can estimate one's volume as follows. Let's assume that I weigh 165 pounds. I first convert my weight to kilograms (1 kg equals 2.2 pounds), so that becomes 75 kg. The density of water is 1 kg/liter. Applying the definition of density, 1 kg/liter equals 75 kg divided by my volume. My volume must thus be 75 liters. In other words, a person's weight in kilograms approximately equals his or her volume in liters. (If someone who weighs 75 kg is a little denser than water, then that person's volume will be a little less than 75 liters.)

I could have performed the same calculation without first converting my weight to kilograms. To do so, I would need to know that the density of water in U.S. customary units is 8.3 pounds per gallon. Applying the definition of density, 8.3 lb/US gal equals 165 lb divided by my volume. My volume in these units is thus 19.8 gallons. Thus, this calculation works either way, in either system of units. In kilograms and liters, the arithmetic is a little easier, since the density of water is a nice, round number: 1 kg per liter. (That figure is also an approximation, since the density of water varies with temperature, but in this case, 1 kg/liter is certainly accurate enough.)

When I try this little volume-estimating exercise on European students, they almost always apply the second method (the density method), and they almost always get a good answer, and quickly. Admittedly, this is hardly a scientific study, but why does it turn out this way? Are these European kids smarter or better educated than their American counterparts? Perhaps, but perhaps not. They do, however, learn one and only one standardized system of measurement and units: the international metric system, known as SI (for *Système international d'unités*). This helps them attain a certain numeracy that is often lacking in their American counterparts. ("Numeracy," as noted earlier, is a relatively new term intended to serve as the numerical equivalent of literacy.) Almost any European could tell you the density of water: 1 kg/liter. In the United States, I would wager not one person in 50 could tell you that water weighs 8.3 pounds/gallon. I would be further willing to bet that an American would be more likely to know the density of water in metric units (1 kg/liter) than in U.S. units (8.3 lb/gal)—even though we have not adopted the metric system in this country. From this, is it a stretch to conclude that the ways that we standardize measurement influence the knowledge that we have? Are some standardized systems of measurement meager and unsatisfactory compared to others? At the very least, it seems clear to me that the ways in which we standardize measurement can make it easier or harder for us to gain certain kinds of knowledge.

Over the years, stretching all the way back to Jefferson's reform of the currency system in 1785, there have been numerous attempts to replace the U.S. Customary System with a decimal system. Nothing has really worked, obviously. The change may be coming upon us gradually, however, driven, ironically enough, by money. (Or perhaps this is only poetic justice!) Multinational corporations that both manufacture and sell their products throughout the world find it increasingly cost-effective to standardize based on one system of units—and that system is the metric system, or SI. Ford Motor Company, for example, now uses only metric sizes on all the fasteners on its cars throughout the world—it's cheaper that way. It may be that commerce and globalization will succeed where legislation and education have failed—and that the United States will finally go metric, if ever so slowly.

CHAPTER 3

THE

RATINGS

GAME

"Overall" Measurements and Rankings

Number Four should have been Number One . . . Thanks, honey.
—Jack Dempsey (dedicating his autobiography to his fourth wife)

In the world of measurement, there is a special kind of number, one that attempts to combine all of the various attributes of something into a single number. These types of numbers become popular because of their convenience. The Dow Jones Industrial Average is a good example. It is a single measurement representing the overall health of the American stock market.

A relatively simple and useful example is the heat index, which indicates how the weather "feels" to a human being. If the temperature is 84°F, it feels worse outside if the humidity is 79% than if it were only 10%. (At 84°F and 79% relative humidity, the heat index is 94, while at 84°F and 10% relative humidity, the heat index is only 79.) The heat index is not a "perfect" measure of how temperature and humidity combine to make us feel. For example, if it is 105°F and 10% relative humidity, the heat index is 100—the same as for 85°F and 85% humidity. These two conditions feel quite a bit different: compare Las Vegas or Phoenix summer weather, at 105/10, with New Orleans or Houston at 85/85). So, the heat index is not perfect, but it is both useful and convenient. Emergency rooms know to expect more heat-related illnesses and disorders when the heat index exceeds, say, 100, and the general public can use the heat index forecast to plan outdoor activities, what to wear, and so on.

Like some but by no means all of these overall measurements, the heat index has some foundation in scientific theory. The theoretical basis for the heat index relates to the mechanisms by which the human body cools itself when it is too hot. We perspire, and the moisture that forms on our skin then evaporates, resulting in the well-known and very powerful "evaporative cooling effect." If you climb out of a swimming pool filled with 80°F water on a day when the air temperature is 95°F with 10% relative humidity, your skin will immediately feel cool—even though the air is much warmer than the water you just got out of. That's evaporative cooling.

The higher the humidity, however, the more difficult it is for our sweat to evaporate. Simply put, since the air is already full of water vapor, there is no place for our perspiration to go. (At 100% relative humidity, our sweat cannot evaporate at all.) The harder it is for our sweat to evaporate, the more difficult it is for us to cool down our bodies, and the hotter we "feel." To return to the initial example, if it is 84°F, we feel much hotter if the relative humidity is 79% than if it is 10%. One often hears people say that they like the desert, because when they are there, they "don't sweat." This is untrue. If it is 100°F and 10% relative humidity, you will sweat quite freely, especially while exercising. However, your perspiration will evaporate rapidly in such hot, dry conditions, so fast that your clothes will not become wet, and it won't seem to you as though you are sweating. That said, you really are sweating—and you can measure it. Weigh yourself before and after going for a run, playing tennis, and so on, in the desert. The weight you lose is almost entirely due to perspiration.

Because measurements like the heat index that combine several factors into one number can be useful, they crop up all over the place. In some cases, these measurements are specifically designed to fulfill a special purpose. In others, they are simply measurements that over time have become popular as shorthand measures for some complex phenomenon.

The Dow Jones Industrial Average (usually referred to simply as "the Dow") is a function of the average price of the stocks of 30 multibillion-dollar American corporations. The Dow is named after Charles Dow, who created his first stock index in 1884. His purpose was to help investors (primarily) and stockbrokers (secondarily) make some sense out of the growing jumble of numbers being created by the ever-increasing number of stocks available to the investor. The Dow Jones Industrial Average (which came along a little later, in 1896) was created for the classic, useful reason for which most overall measurements are created: to help people make sense of a complicated phenomenon by giving them a single number to keep track of.

Over time, the Dow has come to be commonly used as a measure of the overall health of the stock market. During the boom years of the 1990s, it was hugely symbolic when the Dow "finally cracked the 10,000 barrier." Most stock market analysts will tell you that there are better (that is, more representative) single measures of the health of the stock market (the Standard and Poor's 500 average, for example), but the Dow lives on in the minds of many as *the* measure of well-being of the American stock market, and in reality, it does a pretty good job of tracking the overall health of the market.

At least the Dow is real, based as it is on the stock prices of real companies; other such criteria are not. My favorite example is the "stars" assigned to movies by critics. A "four-star" film is supposedly so wonderful that it is simply not to be missed. Three stars is usually pretty good, but anything less than three is generally to be frowned upon (if you really want to see it, save your money and wait for the video—it won't be long). Critics use the star system because so few people are willing to take the time to actually read their reviews of the films. We can get to the bottom line by just looking at how many stars the film got.

One might argue that rating movies with stars is not really measurement at all, because it is so unclear what is really being measured—it all seems rather arbitrary. But it is not quite that simple. To promote motoring and thus increase the sales of its products, the French tire manufacturer Michelin began providing information on hotels and restaurants in the early 1900s. It instituted its now-famous star system beginning in 1926. The Michelin star system is quite rigid and in many ways quantifiable. The inspectors who assign the stars are well trained and discreet. The stars they assign have become over time highly coveted and are worth a great deal of money: if a restaurant or hotel has more stars, it will do more business and can thus charge higher prices. At the very highest levels, the competition is intense. Bernard Loiseau, once the most famous chef in France, committed suicide in 2003 supposedly at least in part because his restaurant, La Côte d'Or, was on the verge of losing the highest Michelin rating of three stars. Michelin is stingy: as of early 2004 there are only 27 three-star restaurants in all of France. Out of the thousands of superb restaurants in a country where fine food is revered with a fervor that borders on the religious, fewer than 150 restaurants have 2 Michelin stars, and only about 400 have even 1 star. When La Côte d'Or received its third star, in 1991, its business increased by 60%.

While assigning "stars" to restaurants or movies might seem trivial or annoying to some, other overall measurements can be downright dangerous. Perhaps none are more so than the various single numbers we use to represent the intelligence of a human being, such as the IQ. How do we combine all the various innate intellectual abilities that all of us possess into one single number?

Reification and Ranking

Others have noticed and written with some dismay about the phenomenon of these overall measurements. Sir Peter Medawar, for example, writes of "the illusion embodied in the ambition to attach a single number valuation to complex quantities." Stephen Jay Gould was not much of a fan of overall measurements, either. Gould speaks of what he calls two "deep-seated fallacies" that can be related to overall measurements. These fallacies are reification and ranking. To reify something that is abstract is to treat it as if it were real and tangible. The term comes from the Latin word *res*, or thing. Gould offers intelligence as a prime example of reification. He notes that intellectual ability is important in our lives, and that we thus desire to describe it. "We therefore give the word 'intelligence' to this wondrously complex and multifaceted set of human capabilities. This shorthand symbol is then reified and intelligence achieves its dubious status as a unitary thing."

Intelligence, as a unitary thing, thus becomes ripe for quantification through an overall measurement, giving rise to the intelligence quotient, or

IQ. This is not necessarily a bad thing in and of itself. But the ways in which overall measurements for intelligence have been abused in the past century represent, collectively, true crimes against humanity.

Gould's second fallacy is ranking, which he describes as "our propensity for ordering complex variation as a gradual and ascending scale." Once intelligence was reified, it was only a matter of time before it was used for rankings. Maybe that ensured that those rankings would be abused.

Ranking American Universities

Gould is not alone in his observation of our love affair with rankings in the United States. Well-respected magazines such as *U.S. News and World Report* and others annually publish overall rankings of American colleges and universities. To produce these rankings *U.S. News* (to use that magazine as an example) collects mountains of data from the schools. Keeping up with the data requirements of such magazines has become a significant task for staff members at many universities these days.

The criteria employed by *U.S. News*, along with the relative weightings of each criterion, are shown in Table 3.1. There are no fundamental scientific principles behind the *U.S. News* rankings, or anyone else's rankings for that matter. You or I could create our own ranking system, using our own criteria and our own weightings.

The subject of university rankings is a personal one for me, since I am a university professor and this is my industry. I am employed by the University of Tulsa, which has for some years now been making a concerted effort to improve its performance in various published rankings. There is nothing wrong with that—I'm certain many other universities have made and continue to make similar efforts. My university's efforts have paid off: in 2003, the University of Tulsa was anointed a "top school" by *U.S. News* for the first time for its overall ranking of 91st. "Top school" sounds right to me. I think my school is an excellent one, and I'm glad to see it get a little recognition. As a small private school (about 4,200 students) in the middle of nowhere (Tulsa, Oklahoma), we face an uphill battle for recognition.

Here's an optimistic way to look at the rankings game, which I believe works for university rankings as well as for both other types of rankings and overall measurements:

Table 3.1

U.S. News and World Report criteria ranking U.S. universities

Individual measurement	Weighting (%)
Peer assessment (reputation)	25
% of students graduating in 6 years or less	16
Freshman retention rate	4
Acceptance rate	1.5
% freshmen in top 10% in high school class	6
Average SAT or ACT scores	7.5
Faculty compensation	7
% faculty with terminal degree	3
% full-time faculty	1
Student/faculty ratio	1
% classes with fewer than 20 students	6
% classes with more than 50 students	2
Average educational expenditures per student	10
% alumni who make financial contributions	5
Graduation rate performance (difference between 6-year rate and predicted rate)	5

Step 1. Whoever makes up the ranking system (*U.S. News* and other magazines for universities, your teacher in a class, your management at work) makes a good-faith effort to select the most important criteria and the most equitable weightings for those criteria and to make those criteria known to those who will be measured.

Step 2. The measuring is done, as accurately and efficiently as possible.

Step 3. Those among the ranked who have done the best job in Step 2 are rewarded with a healthy increase in their ranking. Along with that fine ranking, they can also be relatively certain that they have actually become "better" (that is, that they are now a better university, student, or worker) because they believe in the system created in Step 1.

Step 4. Those among the ranked (the universities, students, or workers), having been made aware of the criteria and weightings, work hard to

improve themselves in as many of the criteria as possible, focusing to
be sure on those with higher weightings.

Step 5. Go back to Step 2 and repeat.

Parents might use the same step-by-step logic when trying to explain to
their child why it is important to achieve a good ranking in his or her high
school class: "The system is a good one," they might say. "Study hard and do
well on your tests, papers, and projects and you will make good grades. Your
class ranking will improve, and you will be rewarded with admission to a
fine college and given a valuable scholarship. More important, you will
have the satisfaction of knowing that you are a well-educated, knowledgeable person, and you will be better prepared for a lifetime of learning."

There are those who view many if not most of these rankings games a
little more cynically. A number of years ago, my university hired a consultant to help it in its efforts to improve its rankings. This consultant had
worked in the "ratings industry" for many years, having been employed at a
well-known journal that annually publishes an influential college rankings
report. I asked the consultant what *one thing* a university could do to
improve its ranking the fastest. That was easy, he said: hire three or four very
well-known people as professors—perhaps a Nobel laureate in literature,
another in chemistry or physics or economics, and a law professor who was
a former U.S. senator. They wouldn't come cheap, but simply hiring them
would create such a splash, in both the academic and the popular media,
that the university's ranking was sure to shoot up, owing to the high weighting placed on reputation (which *U.S. News* calls peer assessment) in university rankings (see Table 3.1). It really didn't matter if these professors
spent only a few weeks a year on campus, or even if they never taught a
class at all.

The reputation category in university rankings is controversial. Reputation is certainly important. Ask any parents who are paying Harvard tuition
for their child—they will tell you in no uncertain terms how important
"reputation" is to them! However, schools with relatively low reputation
rankings relative to their overall ranking (such as my own university, I
hasten to add) often decry the arbitrariness of this criterion. (Those with
high reputations, not surprisingly, tend to think the system works reasonably well.) Reputation is determined through written surveys filled out by

high officials (including the presidents) of the top-ranked universities in the country. It was only human nature, our consultant reasoned, that if those high officials had recently heard the name of a certain university in a favorable light (in connection with the hiring of several prominent faculty members), they would tend to give that school a better reputation score than they otherwise would have.

Well, that was what the consultant said, anyway. I should add that my university's steady rise in the rankings might have resulted from any number of factors, but this get-a-good-ranking-quick scheme suggested by the consultant was never attempted. For one thing, we don't have that kind of money.

I believe there are a lot of valuable data contained in rankings such as those published by *U.S. News*. For example, if I want to see what percentage of a school's alumni contribute money to their alma mater, or if I want to know what percentage of the students at a university were in the top 10% of their high school class, those data are at my fingertips. These data are valuable not only for an academic like me but also for prospective students and their parents. What is the overriding need, then, to combine all this information, somewhat arbitrarily, into an overall measurement of the university? As Stephen Jay Gould (a Harvard professor, by the way) might have said, what is the need to reify and then rank the "goodness" of universities?

I put that question to the above-mentioned consultant, during a meeting between the consultant and about a dozen faculty members on our campus. When I suggested to him that I really didn't see the need to combine all those valuable data into some sort of overall ranking, so why didn't they just publish the data, my fellow faculty members looked at me as if I were from another planet. They seemed to be saying, "Are you nuts? This is the United States of America! We gotta know who's Number One." (This is the same country that ranks the best college football and basketball teams in the country every week by a vote of sports writers, and that similarly ranks young women in the Miss America beauty pageant.)

The consultant was more charitable in his response (after all, we were paying him handsomely for his time). He could see my point, he said, that the data were more valuable than the overall rankings. But I was asking for the impossible. The college rankings issue was a huge moneymaker for the magazine, and its publisher was convinced that it was primarily the "Who's

Number One?" aspect of the issue that really sold it. To publish only the data, and not the overall rankings, would be financial suicide. The magazine publishes both the rankings and the data that go into the rankings. That way, he said, everybody wins.

Fudging the Numbers

Earlier in this chapter, I listed four steps to be undertaken in a good-faith attempt to play the rankings game the right way. Step 1 included "select the most important criteria and the most equitable weightings for those criteria." Let's briefly look at one example of how selecting these criteria and weighting factors can influence the overall ranking—in short, at how to fudge the numbers.

Class size is important at all levels of education, including the university level, and so it is appropriate that this should be part of any overall ranking scheme for universities. (As a teacher, I can vouch for the importance of class size in education, despite what certain nonteachers might tell you.) But how to measure "class size"? What criteria should be used? *U.S. News*, as shown in Table 3.1, measures it two ways. The first is the percentage of classes at a university with 20 or fewer students, and the second is the percentage of classes with 50 or more students. *U.S. News* gives the first of these measures a weighting of 6% in the overall ranking, while the second has a weighting of 2%. Taken together, the 8% overall weighting on "class size" makes it the fourth most important factor in the overall ranking, behind only "reputation" (25%), "6-year graduation rate" (16%), and "expenditures per student" (10%).

Measuring class size this way (that is, using the percentages of classes with fewer than 20 and more than 50 students) is only one way to account for "class size" in an overall assessment of a university. One of the more traditional measurements for "class size" employed by universities over the years is the "student-faculty ratio"—although many have become skeptical of this measure. Universities still love to boast of low ratios on their web sites and elsewhere—say, 11 students for every faculty member. At the same time, many of those same universities are cramming 500 or more students into some of their introductory freshmen and sophomore classes. How can such enormous classes exist when the student-faculty ratio is so low? It is obvious that large numbers of faculty members, especially at research-intensive universities, aren't doing very much classroom teaching. There is nothing inherently wrong with having professors who focus mainly or solely on their research, but at the same time, it is misleading to count those individuals as "faculty" when calculating student-faculty ratios.

A similar measurement often used for class size is, on the surface at least, even more straightforward than the student-faculty ratio. If you want to know something about class size at a university, why not simply sum up the number of students enrolled in all the classes and divide by the total number of classes, thus obtaining the "average class size"? This sounds foolproof. As it turns out, though, while average class size is an excellent measure for an elementary school, it is far less so for a university. At an elementary school, all the students are in class pretty much all day (lunch and recess notwithstanding). They might move from room to room throughout the day, going from math to writing to art to science class, but at most elementary schools you can get a good idea of class size by dividing the total number of students enrolled by the number of classes.

At a university, a "class" is not so easy to define. Many undergraduate students, particularly during their junior and senior years, enroll in independent-study or research "classes" in which there may only be one or two students. From the perspective of the university, these are bona fide classes—the students receive a grade and pay tuition. These one- and two-person classes, however, have the effect of shrinking "average class size" in a way that new university students (and their tuition-paying parents) often find misleading.

For example, let's say you move to a new town and decide to enroll your third-grader in the neighborhood public school, based at least in part on its published "average class size" of 20 students. Your child comes home after the first day of school and reports that there are 500 students in his reading class, and you are outraged. This sort of thing never happens at elementary schools, yet it happens all the time at universities (and yes, parents are frequently outraged).

As a result of all this, many people have became cynical about measurements such as student/faculty ratio and average class size, and I suspect this is why *U.S. News* chose to measure class size differently in its ranking scheme. (As Table 3.1 shows, *U.S. News* does include student/faculty ratio, but only gives it a weighting of 1%—far less than the 8% total for the percentages of classes with fewer than 20 or more than 50 students.) I think *U.S. News*'s motives are good, although the measures its editors chose are still at best a compromise.

A good measurement (for class size or anything else) should be simple. At the same time, it must capture the essence of the thing being measured. Average class size is certainly simple, but it doesn't always capture the essence of the matter—it doesn't always tell us what we want to know (just ask the parent whose kid is in the intro. class with 500 students, even though his university has an average class size of 20).

U.S. News employs two measures that are certainly simple enough. Harvard, which tops the *U.S. News* overall rankings for 2005, has fewer than 20 students in 73% of its classes, while 13% of its classes have more than 50. These measures are simple, but they are far from perfect in terms of representing the realities of education at a particular university for prospective students and their parents. Measuring things this way tends to make the largest schools look better than they really are. In this scheme, a class with 500 students counts the same as a class with 50 students. Big universities often teach lots of small upper-level courses with fewer than 20 students, while herding the freshmen and sophomores into cavernous auditoriums for most of the introductory courses.

From the student's point of view, as well as the professor's, there is a world of difference between classes of 50 and 500 students. (You don't really "teach" a class of 500 so much as you "manage" it.) But, strictly in terms of these ratings, a university would be foolish to divide a class of 500 students into 10 classes of 50 students each. Educationally, it would make sense, but from the perspective of this particular ratings scheme, it would be counterproductive.

We could probably invent a measure or measures that better capture the realities of "class size" at a university. The concept of the "weight average" (not to be confused with a weight*ed* average), as opposed to the more traditional number average, comes to mind as a better measure of average class size. The weight average is used in lots of different scientific fields—for example, in polymer science. To calculate the *weight* average class size, you would sum up the squares of the numbers of students in each class and divide by the total number of students. The *number* average is just the average we're all used to: the total number of students divided by the

Table 3.2

Number- and weight-average class size

	School #1	School #2
Classes # 1–10	20 students	20 students
Class # 11	50 students	500 students
% of classes with < 20 students	91	91
% of classes with > 50 students	9	9
Number-average class size	23	64
Weight-average class size	26	363

number of classes. The weight average gives more "weight" to the bigger classes. Consider the example shown in Table 3.2. Both schools have 10 classes with 20 students each. The 11th class at School #1 has 50 students, while it has 500 students at School #2.

In terms of the *U.S. News* criteria, both schools have the same result: 91% small classes and 9% big classes. The 500-student class at School #2 raises its traditional number average class size to 64—nearly 3 times that of School #1. But the weight average class size at School #2 is nearly 14 times as large as that of School #1. This reflects the much greater likelihood that any individual student is going to get stuck in the class with 500 students. Minimizing that likelihood is what I suspect students and their parents are most interested in, and thus a measure such as weight-average class size has some merit here. The weight average is less familiar, however, as well as being more complicated (involving as it does the squaring of the number of students in each class). Using it risks losing the simplicity inherent in "average class size" or "% of classes with more than 50 students."

All this fuss about class size in universities is intended to make a simple, general point: When developing the criteria with which to rank something, it is important that the criteria embody the things that really matter in the rankings—that really describe the essence of the thing. Think of it this way: If you are going to create a ranking system for something as complex and abstract as "the quality of a university" or "the intelligence of a human being," you can be pretty sure that lots of people are going to look only at your rankings, and not at what underlies them. In that case, what underlies them had better be as true (a mathematician would say "homomorphic") to the thing being ranked as it possibly can.

Life-Cycle Analysis and Utility Analysis

Overall measurements and rankings such as those for universities are often rather arbitrarily constructed. Techniques for creating overall measurements with some basis in theory, mathematical and otherwise, have been developed, however. Two examples of this are life-cycle analysis and utility analysis. Life-cycle analysis is probably used most frequently to create overall measurements related to environmental issues, while utility analysis can be used to help one choose among alternatives for complex, important decisions.

The environment is an area in which overall measurements have seen a lot of use, and for good reason. For example, which is better for the environment, a paper grocery bag or a plastic one? "Paper or plastic?" is a question still asked by cashiers in many grocery stores in the United States. How the individual grocery shopper answers that question depends on many factors. There are probably many shoppers who simply don't care one way or the other. Others may prefer paper or plastic in terms of the ease of carrying the groceries, the probability of bag failure, the ease of bag disposal or reuse in the home, or some other practical, personal consideration. Customers of an environmental bent, though, may indeed be interested in which of these two packaging products is the best for the environment, or "greenest." Evaluating the greenness of something pretty much implies the use of some overall measurement, because so many competing factors are involved. This is where a technique like life-cycle analysis can come in handy.

Looked at strictly from an environmental point of view, paper bags and plastic bags each have their strong and weak points. The paper bag is derived from forestry products (i.e., from trees)—a renewable resource. The plastic bag is made of high-density polyethylene, derived from petroleum, a nonrenewable resource. Score one for the paper bag? Not so fast. One must first, in the case of the paper bag, quantify the nonrenewable resources that go into the planting, growing, and harvesting of the tree, its conversion into paper bags (the paper-making and bag-manufacturing processes), and finally the shipping of the finished bag to the grocery store. There are similar stages in the life of the plastic bag. One must find the crude oil, extract it from the earth, ship it to a refinery (e.g., by pipeline), chemically reform the oil molecules into polyethylene, process the polyethylene into the form of a bag, and ship the bag to the grocery store. At each step in the life cycle of each type of bag, there are environmental costs, all of which can be measured or at least estimated.

In the end, it becomes an accounting problem. Everything, all the materials and all the energy, that flow into and out of the product (the bag) during its "life cycle" can be accounted for on a big spreadsheet. But how do we use all that information to decide which is better (or greener), paper or plastic? We need an overall measurement to help us compare, for example, the bad environmental effects of the emission of chlorine compounds used in the paper-making process with those of the hydrocarbons spilled onto the ground or emitted into the air during the petroleum production, transportation, and refining processes. It's not just a matter of "comparing apples with oranges." We have quite a few different kinds of fruit here, all of which must be combined into one overall number if we wish to know which is better, paper or plastic.

I hope you will not be disappointed to learn that the answer is, "It depends." It depends on all sorts of assumptions, not just in terms of the data that go into that life-cycle spreadsheet, but also in terms of the environmental impacts of the factors behind those data. Society has to decide how important it is to conserve nonrenewable resources, protect the groundwater, lower noise levels, and so on. We have to answer those questions first, that is, we have to quantify our societal preferences for these various environmental impacts, before we can definitively answer the paper or plastic question.

Life-cycle analysis has probably been applied more in the packaging industry than anywhere else. Paper versus plastic is only one of the questions that has been investigated. Styrofoam versus paper "clamshells" for fast-food burgers and aluminum versus steel, glass, or plastic soft-drink containers are among the other packaging-related questions that have been analyzed in this manner. In each case, and in the cases of the application of life-cycle analysis in other industries, questions related to societal preferences keep coming up over and over again. This is where the technique known as utility analysis can, at least in theory, help.

Utility analysis is a formal mathematical-scientific field of inquiry that has as one of its goals, essentially, the creation of single, overall measurements to represent extremely complex factors or sets of data. Application of utility analysis is intended to aid in the making of almost hopelessly complicated yet important decisions, such as where to build a new airport. (Anyone who doubts either the complexity or importance of such a decision should consult the planners of Chicago or Paris, both of which cities are currently in the throes of trying to decide where in their environs to build a third major commercial airport—or whether to build a new airport at all.) In utility analysis, one attempts to take into account all the many factors involved in a decision by first discovering the preferences of the decision-makers related to each of those factors. Let's say your teenager likes to stay up late, but she doesn't like to do housework. You decide to make a deal with her: the earlier she goes to bed, the less time she has to spend each day doing housework. Which do you think she would prefer, going to bed at midnight and doing 2 hours of chores, or going to bed at 10 p.m. and doing 30 minutes of chores? If you spent enough time polling her on this subject, you could eventually quantify her "preferences" with respect to bedtime versus chore time. You could even create a graph with bedtime on one axis and time spent on chores on the other, as shown in Figure 3.1. The curve on this graph would describe all the combinations of bedtime versus chore time to which your daughter is "indifferent." Maybe, in her mind, she has no preference between going to bed at midnight and doing 2 hours of chores versus going to bed at 10 p.m. and doing 30 minutes of chores. Both of those points would then lie on the same "indifference curve" on your graph.

Discovering preferences like this is a big part of using utility analysis in decision-making. Let's get back to the airport location decision. When deciding where to build a new airport, one must consider, among other things, the sizes of the alternative sites, room for future expansion, proximity to highways and other ground transportation, noise issues for nearby residents, various other environmental issues, such as the presence of wetlands or of the habitat of endangered species, national security issues (especially in the wake of September 11, 2001), predictions for future growth patterns in the metropolitan area to be served by the airport, the changing needs of the airlines, and probable advances in airport and airline technology. The preferences of those who will use the airport (travelers and those who will work there) and those who are otherwise affected by the airport

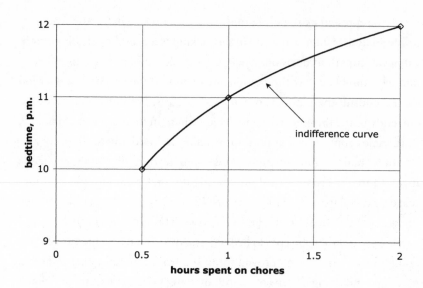

Fig. 3.1. Chore time versus bedtime

(such as those who live nearby) would need to be quantified. What, for example, would people's preferences be for reducing the commuting distance to the airport versus the increased noise level it would create to have the airport closer to town? Because there are so many more people involved, and so many more factors to consider, it's obviously infinitely more complicated than the simple example of bedtime versus chore time.

Utility analysis describes mathematical tools for combining all of these factors for any given alternative into a single overall measurement, the "utility" of that alternative. Utility analysis is complex and difficult to apply, and the results are far from unambiguous, since various assumptions go into the analyses. Nevertheless, it is a means, with at least some basis in theory, by which one can arrive at a single number—an overall measurement— to represent the combined effects of many different qualities.

Would You Like Fries with That?

Overall measurements can be addictive. But despite what we may think about them, they are here to stay. It is thus important to recognize an overall measurement for what it almost always is: an oversimplification of a

Table 3.3

Overall measurements in common use

Overall measurement	Description
Heat index	Combines temperature and humidity into a measurement of how it "feels" outside to a person.
Dow Jones Industrial Average	A function of the average price of 30 common stocks, often taken to be a sign of the overall health of the U.S. stock market (if not the other world markets as well).
Intelligence quotient (IQ)	The result of written or oral tests designed to assign a single number to all the (supposedly) innate and unchangeable intellectual abilities possessed by an individual.
Grade point average (GPA)	In the United States, the average of grades received in academic courses, usually on a scale of 0 to 4. Often used as a single measure for the combination of all a student's abilities and knowledge.
Movie ratings using stars	An ordinal scale of one to four stars (typically), and assigned by movie critics. A shorthand measure for whether a movie is worth seeing.
Soil quality index	Various overall measures (more than one index exists) for the physical properties and "field behavior" of soils, typically for agricultural uses.
University ratings	Various rating systems, the most popular currently being that provided by *U.S. News and World Report,* for the "overall quality" of a college or university. Combines subjective factors (such as reputation) with objective ones (such as graduation rates) into one overall measurement.
Presidential approval rating	The most famous of these is provided by the Gallup Organization. A poll of Americans as to whether they approve or disapprove of the job being done by the president. Used widely for political purposes and as an overall measurement of "how the president is doing."
Gross domestic product (GDP)	An attempt to measure the entire output of the U.S. (or another) economy in the marketplace of goods and services. Often adjusted for changes in prices to give the "real" GDP, and also often expressed as dollars per capita to give a measure of the "standard of living."
Quarterback ratings	An overall measure of the performance of a quarterback in the National Football League. Combines passes thrown, passes completed, yards gained per pass, touchdowns thrown, and interceptions thrown into one overall measure. The current formula has been in use by the NFL since 1973.

typically complex situation. We should all resist the temptation to use overall measurements as a substitute for the careful study of an important issue. But that is easier said than done. Getting people to stop relying on overall measurements is like asking them to swear off fast food. Fast food is popular because it's so easy. Just as Americans are getting fatter and fatter on a steady diet of fast food, we risk getting intellectually soft on a steady diet of overall measurements.

It doesn't have to be that way, and it isn't always. Any stockbroker could tell you how the Dow did today, but he is also smart enough to know that the Dow is only an indicator, and that individual investment decisions should be made based on the specific market information that is relevant to the investment in question. Likewise, a university should evaluate the SAT scores of its applicants, along with their high school GPA, but one hopes that they will also consider a variety of other factors. Numerical factors such as high school class rank are also important, and so are things more difficult to quantify related to activities outside the classroom—the kinds of things the coach of a sports team might call the "intangibles." But just as those intangibles are disappearing from the sporting world, as we measure more and more aspects of athletic performance (see Chapter 5), so they seem to be disappearing from these other areas as well.

Some overall measurements are given in Table 3.3 (see page 53), along with a short description of what each one is and how it is often used.

CHAPTER 4

MEASUREMENT

IN

BUSINESS

What Gets Measured Gets Done

> *Not everything that can be counted counts,*
> *and not everything that counts can be counted.*
> —From a sign that once hung in
> Albert Einstein's study

Trusting the Numbers

Quantification and measurement in the business world are such a part of daily life that they have become the source of countless jokes. Saint Ferdinand may be the patron saint of all engineers, but his secular equivalent would probably be the cartoon character Dilbert. One day, Dilbert complains to his company's purchasing guy because his request for a new computer hasn't been approved. The purchasing guy retorts that this is because Dilbert didn't quantify the new computer's value to the company. Exasperated, Dilbert responds that he can't quantify the value of every item he needs to do his job. For example, why does the company even bother to give him a chair to sit on? He can't quantify the value of that, either! Later, we find Dilbert sitting on the floor of his chair-less cubicle, lamenting, "This is just one more reason why it stinks to be me."

Most of us who work in American industry can identify with Dilbert. But it is possible to quantify the value of Dilbert's new computer to his company, or even of his chair. Sometimes, doing so might even be a good idea! Modern business, in Dilbert's world or in any other industry, is ruled by numbers. How did things get this way? It's been a long, gradual process. Business and arithmetic are naturally linked, just as say, geometry and land surveying are, and have been for a long time. The use of clay tokens as money dates back about 10,000 years. But the origins of the modern numbers-driven world of business can be traced to secular attitudes that were gaining influence in sixteenth-century Europe. In contrast to other parts of the world, where older religious values continued to hold sway, in Europe at that time there was a growing recognition of "utility as self-justifying," the business historian Thomas Cochran notes. Utility, in this case, is the pleasure or satisfaction that one gets from consuming products or services. Providing utility can thus be thought of as the goal of modern business. In the Europe of the sixteenth century, the spread of this concept (at the expense of the influence of the Church) was rapid and profound. When the American colonies were established, this new way of thinking was part of the heritage that came across the Atlantic Ocean with the settlers. Their descendants grew up with this way of thinking, and American business grew up with it as well.

If the goal of business is to provide utility to the customer, it's easy to see why business is such a numbers game. In my own industry, higher education, we have the rankings race, as discussed in Chapter 3. To improve its ranking in *U.S. News* and elsewhere (and thus improve our utility and attract more and better students), my university (along with most of its rivals) attempts to "continuously improve" its faculty/student ratio, the size of its endowment, the percentage of alumni who donate money to the university, average class size, the acceptance rate of its applicants, the standardized test scores of its students, their graduation rates, and so on. Even seemingly qualitative aspects of a university such as its reputation and the prestige of its faculty are measured—perhaps imprecisely or even arbitrarily or unfairly, but nonetheless they are measured. Thus, all these things can be monitored by a school's administration, which can put in place programs to attempt to improve them—programs whose effectiveness can thus also be measured. A university, or any other business, measures things like this and disseminates the results to those who may be able to influence them in the future (in the case of a university, to the faculty, the staff, the students, and the alumni) in order to get people to focus their efforts on improving these measures.

This is somewhat analogous to the generic steps described in Chapter 3 for ranking things: Step 1 is to decide what to measure: businesses want to measure those things that are of the most importance to the organization. That is, those things that are the best indicators of whether the business is making progress toward its goals. Step 2 is to do the measuring, both accurately and efficiently. Step 3 is to let everyone who matters know the results of the measurements. Step 4 is to go about improving performance in each measured area. Step 5 is to reevaluate the results of Step 1, and then go on to Step 2 and continue, in an endless loop. It's hard to argue with this process in theory. If it's done right, it really should quickly and efficiently improve a business. This is because "what gets measured gets done."

What Gets Measured Gets Done

The management consultant Tom Peters became famous in the 1980s for his books *In Search of Excellence* and *A Passion for Excellence.* What some found refreshing about Peters's approach was his "renewed emphasis on the qualitative aspects of business—for example, on people, customer satisfaction, nurturing of unruly champions, and managing by walking around." However much Peters believes in the importance of these qualitative aspects, he at the same time describes himself as a "closet quantifier." He claims that the best management advice he ever heard "is the old saw, 'What gets measured gets done.'"

As employees, we focus our daily efforts on those of our work goals that are measured. A salesperson, for example, typically focuses more on "monthly sales" than on, say, "customer satisfaction," because sales figures are always measured (they are also easy to measure), whereas customer satisfaction is less often measured (and is more difficult to measure). Modern businesses have found that if they want to increase customer satisfaction, they have to take the trouble to measure it regularly and objectively, to make the results known to their sales staff, managers, technicians, and other employees, and to reward those who perform well with respect to this measure. Being a good manager, in these terms, is relatively simple; you just have to design and implement a good and proper program of measurement for your employees.

"What gets measured gets done" is true in areas besides business, too. When I was a student, it was easy to decide what to work on in the evenings after class. I spent my time on the things that were going to be measured! If I had two homework assignments, one that was to be graded and one that would not, I spent my time on the one that was going to be graded. Even though I would eventually be tested (and thus measured) on both subjects, in the short term I worked on the assignment that was going to be, in the short term, measured. I rarely read my textbooks before going to class, since that almost never gets measured in engineering school. In law school, by contrast, most students read for class, because they will be arbitrarily called on to discuss the readings, and their performance in those discussions will then be graded by the professor. Now that I am a teacher, it is relatively easy to get my students to focus on whatever I think is important for them to learn. I simply make sure they know that those things are going to be measured and measured frequently.

The black-and-white certainty that many of us feel as students relates, I believe, to the prevalence of measurement in our schools and to the what-gets-measured-gets-done mentality. Later in life, after our student days are over, some of that comforting certainty is gone. Instead of just focusing, as we did as students, on what is being measured, we are forced (as parents, business persons, teachers, doctors, and so on) to decide ourselves about what is important—about what *will be* measured. That is much harder to do.

Beyond the ratings game for the university as a whole that was described in Chapter 3, my own career as an individual professor is measured in many ways. (I'm going to pick my own profession since I know it best.) Professors at just about any university are charged with "achieving excellence" in three areas: teaching, research, and service to our profession. (The relative importance placed on the three areas varies with the university.) My own progress in each of these areas can be measured to a greater or lesser extent. Early in my career, I was granted tenure by my university as a result of having achieved enough success—most of it clearly measurable—in these key areas of measurement to merit a lifelong appointment in my department. My colleagues and I, tenured or not, continue to be measured as teachers, researchers, and service providers.

As a teacher, I am measured through my students' evaluations of my courses. At my school, these evaluations are both numerical (rated on a scale of 1 to 5) and written. The numerical evaluations are probably given more importance than the written ones; they are (or at least they appear to be) more objective, they require less effort to review, and they can be averaged to create overall measurements and then ranked. Thus, when I see a stack of papers sitting on my desk waiting to be graded, I remember that, very soon, the students who turned in those papers will be asked to use a no. 2 pencil (or, increasingly, a computer mouse) to fill in a circle numbered from 1 to 5 corresponding to their evaluation of my performance relative to "Returns graded materials in a timely manner." My teaching performance can also be measured by the number of awards received, the number of new courses taught, the popularity of my elective courses, and so on. All of these are measurable.

As a researcher, my excellence or lack thereof is measured by the number of high quality scholarly articles I have published ("quality" is measured in part by the reputation of the journal in which the articles are published), the dollar amount of my research contracts and grants, the number of research proposals I have written and to whom, the number of awards I have received for papers and presentations, the number of graduate students for whom I serve as research advisor, and so on, to name only a few of the criteria. My service to my profession can be similarly measured by quantifying my participation on committees inside and outside the university, by my services in editing and refereeing the scholarly work of others in my field, and so on.

Every industry has its analogous measurements. If a company builds cars, it measures its sales, its market share, an almost infinite variety of parameters related to its costs, from labor to materials to shipping, advertising, and so on, the amount of time, in labor-hours, it takes to build a car, numerous parameters related to quality, the rate of customer complaints classified by their nature, and on and on. All of these measurements are monitored for changes over time and are endlessly compared with those of rival companies. U.S. carmakers, for example, can never seem to catch up to their Japanese counterparts in terms of the number of labor-hours required to assemble a car—an important measure of efficiency in manufacturing. A June 2005 report by Harbour Consulting on North American automobile assembly plants concludes that Toyota is the most efficient carmaker in North America, requiring an average of 27.9 labor-hours to manufacture a car, followed by Nissan (29.4 hours), Honda (32.0), GM (34.3), DaimlerChrysler (35.9), and Ford (37.0).

Several years ago I was in a General Motors automobile assembly plant where I was shown a large, sparsely furnished, and antiseptically clean room. Distributed about the room in carefully organized patterns were the parts, and the parts of the parts, of a brand-new Toyota Camry. This unfortunate Camry was in the process of being completely dissected and measured to the nth degree. The number, size, and uniformity of the spot welds on the steel frame and the distance between those welds was measured. The density, thickness, and weight of the foam seat cushions, the size, type, and number of mechanical fasteners under the dashboard, the weave and thread count of the upholstery fabric, the amount of adhesive used to bond the windshield—all these and a mind-numbing total of other measurements were being made. The labor required to dissect, measure, and analyze the data from this one Camry could itself be measured in the hundreds if not thousands of hours. All of the data generated in this "reverse engineering" process were being compared to those of GM's own vehicles and to the vehicles of other competitors—to see in what ways the GM car was better and where Toyota or another company were better, and to thus help plot GM's future course. Manufacturing companies are thus among the most reliable consumers of their competitors' products. Someone in the photocopier manufacturing business once told me, slightly tongue in cheek, that the main reason his company produced its largest copier model (a room-sized, million-dollar machine that more or less sucked whole trees in at one end and spewed out a blizzard of copies at the other) was that, even if no one else bought it, they knew that each of their competitors would in order to reverse engineer it.

You Can Measure Anything

Modern business can be classified, as it is done in a business school's curriculum, into its respective elements such as marketing, accounting, finance, and management. Each of these can be further broken down into various subspecialties. Wherever you look throughout the spectrum of modern business, you find a discipline that depends on and is driven by measurement—by "the numbers." It is not surprising that activities such as accounting and finance would be driven by measurement. But what about the "softer" aspects of business, things like management and marketing? There, too, we find that everything is by the numbers.

For those of us who are not managers or business professionals (myself included), it might seem that in business, management would be the last refuge of the qualitative. What do we manage, after all? It's not *what*, it's *who*—we manage people. The good manager has to understand her employees. She has to "get inside their heads" and know what motivates them—to find those keys and turn them constantly so as to get the best out of her people. This sounds more like a job for Sigmund Freud than one for Isaac Newton. Isn't there something about this that is inherently nonquantifiable? Perhaps, but like all other aspects of business, management is today essentially a measurement game.

Anyone who has worked in a large organization has probably had some experience with management by measurement. Those experiences are not necessarily good ones. I still get the shivers when I think back upon the measurement program that a team of management consultants inflicted upon a group of engineers I worked in during the 1980s. For several weeks, we were required to record our daily activities in fifteen-minute increments. For each quarter-hour period, we had to enter onto a spreadsheet a numerical code corresponding to our principal activity for that time interval. There were codes for making calculations, talking with clients on the phone, meeting with clients in person, searching for vendor data, running experiments in the lab, and so on. The scope of all possible activities was relatively broad—there was even a code for "thinking." (One of my colleagues wryly observed that this category should have been further subdivided into "thinking with eyes open" and "thinking with eyes closed.")

In some ways, what we were doing wasn't so unusual—accountants and lawyers often bill their clients in fifteen-minute increments. But in this case, my co-workers and I knew what this was all about. Some of us were going to be downsized, and this was management's way of justifying the job cuts. This points out what the management consultant Aubrey Daniels sees as the greatest misuse of measurement as a management tool. When he worked as a clinical psychologist, Daniels never heard a patient complain about "being measured." He believes this is because his psychological patients were convinced that they were being measured in order to be helped. In business, Daniels says, most of us believe that when our management measures us, they are doing so in order to punish us.

Measurement in management, generally speaking, is not especially high-tech. It's mostly just a matter of counting things (sales per month in dollars, quality parts produced per day, customer complaints per week) or judging things. What do we judge? The things we can't count or otherwise measure. Just as we judge movies on a scale of one to four stars, restaurants on a scale of one to three stars, or figure skaters on a scale of one to ten, in management, workers are often judged with respect to aspects of their performance for which other types of measurement do not exist. Daniels says the key to this is to make the process as objective as possible. To do this, it is necessary to link the judging scales to as many specific behaviors as you can. For example, if employees are being judged on a scale of one to ten with respect to "customer service," it is important to be very specific in terms of what constitutes a ten, a nine, and so on. Just as a figure skater gets a specific number of points deducted for touching a hand to the ice, an employee loses points for failing to follow up on a customer complaint in writing within a specified period of time.

A properly developed managerial measurement program has many of the same benefits as a similar program in other fields of endeavor. One of those benefits is the ability to see small changes quickly. A pitching coach in baseball who modifies the throwing motion of a top young prospect wants to know the results of those changes. Is the young phenom's fastball really getting faster? A speed gun provides the answers, in increments much smaller than the coach or another witness—or the pitcher himself—would be able to observe.

The same thing is true in management. The good manager learns to believe in his measurements, and not to be swayed by his emotions or by what somebody *thinks* is going on. For example, several waiters in a restaurant may tell the manager that a recently instituted incentive program to help the restaurant sell more dessert really isn't having much effect, but the manager should know better. It is easy enough to measure how much dessert is being sold.

The modern emphasis on measurement in management is often traced to the work of Frederick Taylor, who wrote *The Principles of Scientific Management* in 1911. In the 1880s Taylor went to work in a steel mill, and his observations of the workers there led him to develop, among other things, what are now known as time-and-motion studies. These techniques involve the precise measurement of the amount of time it takes a worker to perform each individual task and subtask required to do his job. The results allow management to modify individual jobs, processes, and work flow in an organization so as to streamline production and improve productivity. "Taylorism" as Taylor's principles came to be known, has generally negative connotations these days—it conjures up images of some poor, harried laborer slaving away on the assembly line while a cold, unfeeling manager in a white lab coat scrutinizes his every move, stopwatch and clipboard in hand.

Some principles of Taylorism have been discredited, while others in modified form are still used. The point is, management embraced measurement as a fundamental tool beginning with Taylor and continuing to this day. Today, the question in business is not whether to measure, but how to measure. To earn an MBA, students learn to master the quantitative tools of business. They learn how and what to measure in order to "get it done." The dean of a prestigious business school once told me that she loved receiving applications for her school's MBA program from engineers, because she knew with absolute certainty that these students could handle the mathematical rigors of her program with ease, regardless of where they had gone to engineering school. However, she said, she was always a little nervous to admit students who had studied less quantitative subjects as undergraduates (even at the nation's top-ranked schools), because these students were frequently overwhelmed by the math requirements of business school. About a third of all MBAs go to engineers, despite the fact that only about 6% of undergraduate degrees in the United States are in engineering. There is something to be said for the logic, discipline, and determination engendered by a rigorous quantitative education, although I don't think we engineers will be taking over the world of business anytime soon even so.

M-i-c-k-e-y . . .

I've only been to Walt Disney World once, in 1996. It was a nice vacation, but what has stayed with me over the years is not so much the thrills or the sights, but the uncanny, almost eerie way Disney understands the various needs (corporal, spiritual, and so on) of the human being—or, more precisely, of the human being on vacation. From the moment you wake up at a Disney World hotel and throughout your day, Disney seems to have you figured out—they seem to know you better even than your own mother does. Need to get somewhere? Comfortable transportation is available, a few steps away. Hungry or thirsty? Whatever you're in the mood for is right around the corner. Tired? A cool place to kick back is sure to be close by. Feel like shopping? Have they got some deals for you!

Eventually, I grew a little bit annoyed at the thoroughness with which they seemed to have taken me apart and analyzed me. I'd always thought I was this complex, inscrutable guy, but to Disney, I was just another customer to be firmly but gently and ever so thoroughly separated from his money. Even worse than the realization that someone could have me figured out so accurately was that I actually enjoyed the process.

Disney has clearly mastered the elements of marketing, including new product planning, development, pricing, promotion, and distribution better than anyone else in its industry. That most aspects of this are measurement-driven should not be surprising, but let's look at a different example where the innovative use of measurement in marketing is both surprising and extremely effective.

I Shop, Therefore I Am

I thought Steven Spielberg's 2002 film *Minority Report* was rather silly, but there was at least one aspect of it that fascinated me. When a character in the film enters a retail clothing store, retinal scanners in the store quickly identify him, and the "liquid advertisements" on the store's walls are instantly modified to correspond to his tastes and buying habits. He sees video images of clothing in the styles, colors, and sizes that he prefers. The audio portion of these tailor-made advertisements even addresses him by name and asks how he likes a recent purchase he has made.

The scene was obviously intended to remind us of Big Brother, and it was probably inspired by the constant barrage of advertisements we receive by e-mail and when searching the Web. Some (but by no means all) of this computer advertisement is eerily tailored to our individual tastes: amazon.com for example is always trying to sell me books and music by my favorite authors and musicians and is even pretty good at suggesting new artists with whom I am unfamiliar. The main difference between what we are subjected to on our personal computers and what happens in the store in *Minority Report* is the futuristic (not to mention invasive) use of retinal scanning in the film.

But what that movie scene really reminded me of was the innovative work of Paco Underhill. Underhill has applied his training as an anthropologist to "the science of shopping." Bloomingdales, Starbucks, and Blockbuster are his laboratories (as well as the clients for his successful marketing research firm, Envirosell). Early in his career, Underhill worked with the pioneering anthropologist William H. Whyte in New York City on the Project for Public Spaces, which investigated (and continues to investigate) how people use public spaces like parks, plazas, and sidewalks. Underhill built on that work and took it in a different, more profitable direction: the retail store.

The science of anthropology had neglected the very human activity of shopping, according to Underhill, so he took it upon himself to apply science to shopping, or, more precisely, to the shopper. Underhill's goals, to be sure, are more than purely scientific. That is, he wants to do more than just understand how people shop and what makes them buy. He makes a living doing this, after all. He has to help companies apply the things he's learned about shopping to sell more, sell faster, and sell more efficiently. How does he do this? He watches people shop, and he measures things— lots of things. He can tell you how many males who take jeans into the fitting room will buy them (65%), compared to females (25%); how many shoppers in a mall housewares store use baskets (8%); and how many of those who take baskets actually buy something (75%), compared to those who shop without baskets (34%). On this basis, Underhill then suggests ways to his clients to increase the number of shoppers who use baskets.

What Paco Underhill has done for shopping is in the best sense a classic measurement story. Before he quantified shopping, store managers and marketing executives were largely in the dark regarding the crucial link between products and profits that shopping represents. Their knowledge of the shopper and his habits was truly meager and unsatisfactory. But just as Wallace Sabine quantified what makes a music hall sound good, Underhill quantified (and continues to quantify) what makes us pick something off the shelf and pay for it.

CHAPTER 5

GAMES

OF

INCHES

Sports and Measurement

Ninety percent of this game is half mental.
—Yogi Berra

A Rainy Day in September 2003

During an interminable rain delay at the U.S. Open tennis tournament on September 1, 2003, the television producers were obviously getting desperate. Having replayed all of the highlights of the previous several days' play again and again (not to mention a depressing sampling of the lowlights), they were finally forced to dig deep into their video archives and drag out a real golden oldie: the famous 1991 match between a then 39-year-old Jimmy Connors and young Aaron Krickstein. At 7–all in the second-set tiebreaker, Connors blasts a shot that lands near the sideline. The linesman calls the ball "good." Krickstein protests, and the umpire quickly overrules the call, awarding the point to Krickstein. In a flash, Connors goes absolutely ballistic, screaming four-letter words at the umpire, impugning his character, demanding he step down, and so on. Connors even launches into his now-infamous rant that goes something like, "Here I am, just a working-class stiff, getting on in years. I'm out here doing my best, working my rear end off trying to make a buck, and now here you come, you idiot of an umpire, taking food out of my babies' mouths." A bit over the top, but it all made for great TV drama, even twelve years after the fact.

Had this match been played in 2003, instead of 1991, here's what would have happened next: the TV coverage would have featured one of two video replays of the shot in question. Which replay the viewers saw would have depended on whether the match was being broadcast by CBS or by its rival USA Network. CBS, utilizing the DuPont "Mac Cam," would have shown an ultra-slow-motion high-resolution view of the ball landing near the line, such that it would have been possible to know, visually and within perhaps a few millimeters, whether any part of the ball had actually touched the sideline when it bounced. The USA Network replay, employing "Hawk-eye" replay technology, would have been computer-generated. Using an ingenious four-camera system connected to a computer via sophisticated software, the flight of the ball would have been measured in real time and its precise path calculated by the computer, which would then have plotted and displayed graphically on the TV screen the spot where the ball touched the court to within an accuracy of (it is claimed) less than one millimeter. Hawkeye (and other similar systems in tennis and now baseball) operates somewhat like a video version of a global positioning system for sports. Millions of tennis fans around the world would have known, with near certainty and almost instantly, whether that ball was in or out.

However, this match was videotaped in 1991. Here's what did happen: A not-so-slow-motion replay of Connors's shot shows a rather fuzzy image of something resembling a tennis ball landing somewhere near (on?) the outside edge of a blurry white line. "Wow, that was close," is about all you can say.

If you asked him today, Connors would probably tell you he still believes that shot was good (he finally did win the match, by the way). Krickstein would be just as likely to tell you it was out. There is no way of knowing for sure. Had this match been played in the U.S. Open in 2003, there would be no point in asking the players for their opinions on the subject. It would be a little liking asking an Olympic sprinter whether she thinks she really did run the 100 meters in precisely 10.48 seconds or not. Who cares what she thinks about that? The clock will tell us for sure what her time was. Today, whether Connors's shot was in or out would be transparently obvious: knowable to all. It would be measured right there for us on our TV screens.

In the 1970s, the fiery Romanian tennis star Ilie Nastase gained notoriety for rubbing out ball marks on soft clay courts with his feet before the chair umpire had time to scamper over and closely inspect them. Nastase's antics today would be even more ridiculous than they were then—the computer and slow motion camera would already have all the data they needed—no need for anyone to interpret a ball mark!

It seems to me, a longtime observer and avid player and fan of the game, that professional tennis players are less likely to violently protest line calls today than they were a generation ago. Perhaps today's best players are more genteel—there are fewer boors like Jimmy Connors. Or perhaps they simply realize how ridiculous they will look if they are seen protesting a decision that millions of people know with absolute certainty was correct.

The umpires of today's tennis matches cannot yet avail themselves of these exceptional new technologies. But all the evidence suggests that these technologies are so precise, so good (and bound to keep getting better), that their eventual inclusion in the game is inevitable. It's just a matter of how and when.

During the 2003 U.S. Open, players began looking up at the TV commentator's booth right after a questionable line call—in the hopes that John McEnroe or one of the other announcers would signal to them whether the ball was in or out. The players were well aware that the announcers possessed the results of the precise measurements (Mac Cam or Hawkeye) and knew the right answer for sure.

For purposes of refereeing matches, Hawkeye (the computer-generated measurement) possesses certain advantages. It is probably more accurate, less subjective, and one system covers all the lines of a tennis court. But Hawkeye's advantages do not stop there. This system has tremendous potential for measuring and thus helping to improve the performance of players—I would be very surprised if it's not already being used in that way. During matches broadcast by USA Network, images—maps of the court— would frequently be shown of the multiple locations (the spots where the balls bounced on the court) of the first and second serves of the players, and even of the points in space above the court where the players' rackets struck the balls on return-of-serve. This is clearly an excellent performance measurement tool for the player and his coach (not to mention quite interesting for the TV tennis fan). It is one thing for a coach to tell his player, for example, "You need to get in closer when you return the second serve. You're standing too far back!" I've been on both sides (coach and player) of that piece of advice. It would be so much more effective if the coach could show the player a map of all his second-serve returns: "Look at this chart! On 19 out of 20 second serves, you made contact with the ball behind the baseline—you gotta get in closer!" The same thing would apply when scouting an opponent for an upcoming match.

The above example is intended to highlight the tremendous changes taking place in all sports today, not just tennis. Sports are truly being revolutionized by the innovative integration of measurement technology in training—on the field of play, in the gym, at the training table, and everywhere else. It is in some ways analogous to the revolution that preceded it in the business world, described earlier in this book.

The Games We Measure

Measurement in sports has a long history. The Olympic Games in ancient Greece included sports whose results are measurable, such as the discus, long jump, and javelin. But a sport whose objective is quantifiable is one thing; it is quite another thing altogether to use measurement to control all the various aspects of how the athlete gets to that objective. Athletes, coaches, trainers, and sports executives, in short, everyone who works in sports (right down to the peanut vendor)—all of them use measurement to generate knowledge and achieve their goals.

Things weren't always this way. Many former high school football players of a certain age recall, with a fondness that grows with the years, the image of their coach, clipboard and stopwatch in hand, timing his players in the 40-yard dash. Periodic measurements of this type were indicators of improvement for both coach and player. But sports in those days weren't driven by the numbers the way they are today. Today's elite (and, increasingly, not-so-elite) athletes are quantified to a truly astonishing degree. The Tour de France bicycling hero Lance Armstrong exemplifies this trend. Armstrong is said to weigh every meal—to quantify every food item that he puts in his mouth. Aerobic capacity, percent body fat, and aerodynamic drag coefficient on the bike are just a few of the many ways in which he is routinely measured. Two racing helmets look identical—yet one, due to its improved aerodynamics, shaves 23 seconds off Lance's simulated time in a 19-kilometer time trial—as determined through wind-tunnel testing. Similarly, raising the saddle on Lance's new bike a mere 5 millimeters (0.2 inches) knocks 96 seconds off the same simulated time trial due to the combined improvements in pedaling efficiency and aerodynamics.

Armstrong and his coach Chris Carmichael have elevated the measurement of training and of performance to a level that seems obsessive to many. Yet Lance's results seem to justify this approach—he has won cycling's most prestigious event, the Tour de France, a record seven times in a row (and that's after recovering from testicular cancer that eventually spread to his brain). Armstrong is by no means alone in his slavish devotion to measurement. Top athletes in every sport now measure parameters that were left unquantified just a few years ago.

Another example of the ubiquity of measurement in sports comes from the world of car racing. Racecars these days are instrumented "to the teeth." Pretty much everything a driver does in his racecar is being measured by some instrument or another. Racing teams from NASCAR to Formula One to Indy Car routinely hire the best engineers they can find these days—and they can find some pretty good ones, because in this lucrative business, they can pay them extremely well. One role that these engineers play on these racing teams is in the instrumenting, data acquisition, and data reduction associated with car and driver.

When you or I drive down the street (I'm not a racecar driver, and I'm assuming you're not either) and we see that the road ahead curves, we use our driving experience to decide how much we need to slow down to safely negotiate the curve. In other words, we know our speed (from the dashboard speedometer), can judge the distance to the curve visually, and can also visually judge the sharpness of the curve, see the road signage, and evaluate the road condition (dry, wet, icy, and so on). While doing all this, and as we approach the curve, we back off the accelerator, in plenty of time for the curve. If necessary, we then apply the brakes—and we successfully negotiate the curve.

This isn't at all how things work in a racecar. As one racing engineer explained it to me, there are only two positions for the brake and the accelerator in a racecar: you're either accelerating 100% (pedal to the metal), or you're braking 100%. If you want to get around the course as fast as possible, there's no middle ground. The racecar driver approaches the same curve I described above with the gas all the way to the floor. At the last possible instant, in order to keep from overshooting the curve and going off the road, he slams on the brakes just long enough to reduce his speed sufficiently to successfully negotiate the curve, and then it's back on the gas, 100%. Any other approach to the curve guarantees that the car will lose time. Other than measuring the time it takes the car to go around the curve (which we've been able to do for as long as cars have been around), what can we measure about the driver's performance in this case? The time it takes the car to go around the curve is a function not just of the driver but also of the car. We need measurements that quantify the driver's behavior independent of the car.

These days, it's pretty easy to measure the position of the accelerator pedal and the brake pedal in a racecar, as a function of both the car's position on the race course and its speed. All you need are the right instruments and a computerized data acquisition system. After the race, or the training session, then, the driver can review the data and perhaps see that, in fact, his foot did come off the gas well before it was necessary (and before he *thought* he decelerated) going into that curve, and that precious milliseconds were lost during that particular lap.

A Business Approach to Sports Strategy in Baseball

There is a growing trend in sports such as baseball to "stick by the numbers" when evaluating performance, as opposed to relying on the old-fashioned intuitive approach. An example of this trend is the growing popularity of what is known as "sabermetrics" in baseball. (The unusual name "sabermetrics" is derived from SABR, the Society for American Baseball Research, which had about 7,000 members in 2002.)

Baseball, as even the average fan knows, has a long history of what might be termed performance quantification. Almost as long as the sport has been around (it began in the mid 1800s), its adherents have been measuring the performance of its players. The father of baseball statistics, Henry Chadwick, developed the modern box score and introduced measures such as the earned-run average and the batting average well before 1900. Although Chadwick did not play baseball himself, he is nonetheless a member of the sport's Hall of Fame. The statistics Chadwick developed and promoted are the most popular and remain, at least in the eyes of most fans, the most important measures of baseball ability.

Throughout the twentieth century, however, occasional assaults were made on the primacy of Chadwick's measures by various numbers-oriented baseball people—who believed there were better ways to measure baseball results. In the 1950s and 1960s, Earnshaw Cook, a mechanical engineering professor at Johns Hopkins University, gained some notoriety for his radical ideas about baseball strategy and statistical analyses. But the person who truly brought baseball analysis into the modern era, today's counterpart to Henry Chadwick, was Bill James, the father of sabermetrics. James describes sabermetrics very simply as "the search for objective knowledge about baseball." James realized that many of the measures developed by Chadwick did a poor job of objectively evaluating baseball ability. Runs batted in, for example, is a combination of skill and luck. To bat in a lot of runs, a player needs batting skill, but he also needs to be lucky enough to have the batters who preceded him get on base and in position to be batted in. James has devoted his career to developing and popularizing more objective and incisive measures of baseball skill.

Just as a modern businessman now believes he can and should choose what stock to purchase, what company to buy, or what product to develop based strictly on an analysis of the business fundamentals, baseball sabermetricians believe a team can and should decide who should pinch-hit, which relief pitcher should face which opposing batter, and which free agent should be offered a multiyear contract based entirely on an analysis of the appropriate baseball fundamentals, or sabermetrics.

Those coaches, general managers, and others who make use of sabermetrics rely on the sober analysis of carefully developed measurements when evaluating players and developing game strategies. These carefully selected statistics are likely to be those developed by Bill James and not Henry Chadwick. The array of statistics in the sabermetrician's arsenal is truly bewildering. But the basic approach to developing sabermetric measurements is refreshing. Instead of just using measurements such as batting average, because "we've always done it that way," sabermetricians stepped back and asked themselves what measurements of a player's performance are the best predictors of whether his team will win games. Winning games is what baseball is all about. Sabermetrics has used this simple approach to show, for example, that a measurement known as "runs created" is a much better measure of a player's offensive performance than the batting average. Runs created is found by multiplying a player's on-base percentage by the total bases he achieved. Runs created is not a statistic you are likely to find in the sports page of your local newspaper—but you can bet the management of your favorite team is quite familiar with it.

In his fascinating book *MoneyBall*, Michael Lewis tells the story of Bill James and others involved in baseball's measurement revolution. *Money-Ball* is mainly the story of the Oakland A's of the late 1990s and early 2000s— the first major league team to devote itself to the new measurement principles of Bill James and the others—including both on-the-field decisions and those made in the front office. As a result of this new approach, the A's had several consecutive seasons in which their performance generally exceeded expectations. That is, they won a lot of games despite a lack of established star players and with one of the lowest payrolls in the major leagues. This turned a lot of heads in the baseball world and gave a stamp of legitimacy to the A's approach. The A's devotion to quantification in the management of their team is absolute and extends all the way down to the scouting of high school and college players.

Can you imagine making a major investment—buying a business costing millions, for example—based on how good-looking the executives, office staff, and factory workers of the company were? To a large degree, that's how baseball players have been chosen for a long, long time. Baseball scouts have often been so blinded by the physical appearance (good or bad) of a young prospect that they were unable to objectively evaluate that player's true potential to help a team win ball games. Just about everyone evaluates people based on how they look, but in baseball, physical appearance has nothing to do with winning and losing. Baseball, now more than ever, is a numbers game, and laptop computers are replacing the baseball scouts who traditionally traveled around the country, watched players from the bleachers, and reported on their games. The sabermetricians keep coming up with more and better numbers to evaluate performance, and the computer crunches out those numbers with ease. That young prospect gets on base 60% of the time and walks five times for every one time he strikes out. So it doesn't matter if he's got a big butt—the numbers say he's a winner.

Not every baseball decision-maker is convinced by sabermetrics yet. But in baseball (and other sports), the time-honored tradition of making decisions based on an unquantifiable "hunch" or a "gut feel" (or by someone's physical appearance) instead of what the numbers say is slowly giving way to the precise quantitative approach, just as it did in business years earlier.

Clutch Performers and Hot Hands— A Few Sports Myths Demystified

The world of sports provides society with more than its fair share of clichés. Many of these have a measurement theme. The athlete who always tries his hardest and thus "gives 110%" is welcome on any team. The difference between winning and losing is so small in most sports that we always seem to be playing "a game of inches." Coaches are forever searching for the right "chemistry" for their team, implying that this is a scientific enterprise and that, somewhere, there is a recipe that can be followed. And when those same coaches wax eloquent about the "intangibles" inherent in their sports, they are following a time-honored tradition. The intangibles in any sport are those things that cannot (yet) be measured. Just as they have in business, the intangibles in sports are disappearing. They are being replaced by cold, hard measurements.

For example, "clutch performers," players who can be counted on to succeed at the most crucial stages of a sporting event, are commonly considered to be among the rarest of all athletes. Yet research in baseball and basketball has shown that, statistically, clutch performance is probably a myth. That is, the athlete you most want to rely on during the crucial stages of a sporting event is in all likelihood the athlete statistically most likely to succeed at any other stage of that contest.

In addition, it is well known in sports like baseball and basketball that some athletes will frequently get a "hot hand" or, at the other extreme, go into a slump at various times during a long season. Statistical analysis of these events shows that periods of "hot" or "cold" performance by athletes (for example, hitters in baseball or shooters in basketball) is likely to be just a reflection of the natural variability in "coin tossing." A statistician will tell you that if you toss a fair coin 100 times, the odds are very strong that there will be at least one streak of six consecutive heads or tails. A fair coin that has just landed six consecutive heads is no more likely to come up heads the next time it is tossed than any other fair coin, and it appears that the recent history of a baseball hitter or basketball shooter is likewise a poor predictor of the near future for that performer. This is the conclusion of researchers who studied and statistically analyzed the performances of Charles Barkley and other NBA stars in the 1980s.

Measuring Yesterday's Performers and Comparing Them to Today's

One of the favorite topics of discussion of sports fans of a certain age involves comparing the relative merits of former versus current stars. How would Willie Mays have fared batting against Pedro Martinez? Could Shaquille O'Neal have taken it to the hoop over Bill Russell in his prime? Would Johnny Unitas have been able to pick apart a modern defense in the National Football League the way he did back in the 1960s? Since none of these epic confrontations will ever take place, except in our imaginations, we shall never know for sure what would have happened, and these questions thus make for great conversation and help many a sports fan while away a long winter evening. Those of us who are a little bit more "experienced," shall we say, tend to believe that were they to have gone mano a mano with today's overpriced, overrated, egomaniacal athletes, the heroes of our youth would have been more than able to hold their own. Wait just a minute, younger fans respond. Today's athletes are bigger, faster, stronger, better trained, they eat better, and they just simply *are* better. They would mop the floor with the stars of yesteryear. No contest!

What does measurement have to say about the comparison of yesterday's athlete to today's? It is simpler to compare sports where there is an objective, quantitative standard, such as running, swimming, weight lifting, or the like. Even the most casual sports fan knows that records in such sports have relentlessly improved over time.

The men's world record for the 1,500 meter run (the so-called metric mile), for example, has gone from 3:56.8 in 1912 (the first recorded record) to the current record of 3:26.0, set by the Moroccan Hicham El Gerrouj in Rome in 1998. The 30.8-second reduction in that record since 1912 repre-

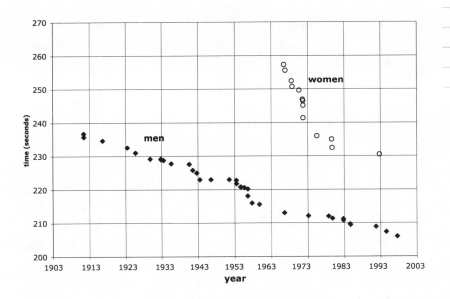

Fig. 5.1. Men's and women's 1,500 meter records

sents a 15% improvement. If you plot the record times versus the dates on which the records were set, the men's data are relatively linear, as shown in Figure 5.1. Thus, the record is steadily improving at a rate of about 1 second every 2.8 years. (The women's data, discussed later in this chapter, are somewhat different.)

Similar improvements are seen in just about any other athletic event where the results are measured by time, distance, and so on, for both men and women. By this measure, it is clear that today's athlete is far better than his or her predecessors. Many reasons are frequently cited for this. Today's athletes really are bigger and stronger, they eat better, they train harder and smarter, and they use better equipment. They are also—big surprise!—more measurement-driven in all that they do to improve their performances.

There is at first glance no obvious reason to think that the same thing shouldn't apply in the case of those sports where superiority is measured more subjectively. Basketball is a good example. Most sports fans have seen scratchy old videos of those skinny, short basketball stars of the 1950s. Just what sport were those guys playing, anyway? Tiddlywinks? It looks like they are moving in slow motion! Shaquille O'Neal, Michael Jordan, and their contemporaries would hardly break a sweat making mincemeat out of the best of those old-fashioned teams. Anyone who suggested otherwise would get laughed right out of the sports bar, right?

Well, let me suggest, gently, that things may not be quite that simple. I must resort to my own favorite sport, tennis, for the following example. Tennis, like basketball and unlike running, is a sport with no objective, quantifiable standard for greatness. Consider the case of the Australian tennis star Ken Rosewall. Rosewall's name is on a short list of the all-time great tennis players who never won the Wimbledon championship. He was excluded from playing at Wimbledon during much of his athletic prime, from 1958 to 1968, when he was a professional player, and Wimbledon, along with the other major tennis tournaments, was only for amateurs. But he did manage to reach the final round of Wimbledon on two occasions, as an amateur in 1954, when he was 19, and again as a professional in 1974, when he was 39. In 1954, he lost to the Czech Jaroslav Drobny, who was then 33 years old. In 1974, Rosewall lost to the 22-year-old Jimmy Connors. Then, 17 years later Connors, in 1991 at the age of 39, reached the semifinal round of the U.S. Open (having famously and noisily disposed of the unfortunate Aaron Krickstein along the way). At the 1991 U.S. Open, Connors was competing in the same event with the defending champion, Pete Sampras, who finally ended his career by winning the U.S. Open in 2002.

Do you see where this argument is going? Jaroslav Drobny played his first Wimbledon championship at the age of 16 in 1938. He finally won the tournament in 1954, beating Rosewall. Rosewall again reached the final in 1974, losing to Connors. Connors came out of retirement to reach the semi-finals of the U.S. Open in 1991, the year after Sampras won his first U.S. Open title. Thus, in 1938, Drobny was a world-class player. He was Wimbledon champion in 1954, beating Rosewall. Rosewall was still good enough at 39 in 1974 to reach the Wimbledon final, losing to Connors, who was good enough at 39 to reach the semis of the U.S. Open in 1991, the year after Sampras won there. Sampras ended his career by winning the U.S. Open one last time in 2002. In terms of tennis greatness, we can get from 1938 to 2002, a span of 64 years, by linking only four players: Drobny to Rosewall to Connors to Sampras. These tennis milestones are superimposed on a plot of men's marathon record times in Figure 5.2.

Between 1938 and 2002, the men's record for the marathon improved by

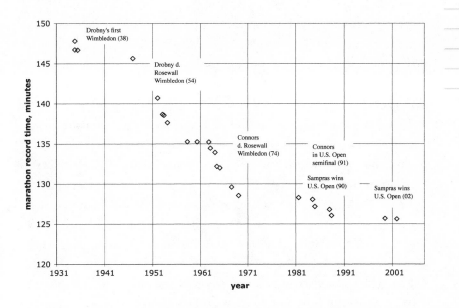

Fig. 5.2. Men's marathon records and tennis milestones

more than 21 minutes, or over 14%. This means that the 2002 record holder would cross the finish line with a lead of about 3.8 *miles* over the slowpoke who held the record in 1938. (The marathon is a 26.2 mile race.) For argument's sake, let's assume that the best tennis player in the world today is also about 14% better than the best player was in 1938. As the 3.8 mile example shows, that is one heck of a lot.

How then would the Drobny of 1938 have fared against the Sampras of 2002? If you watch a videotape of an old tennis match (and it really doesn't have to be that old), the contrast with today's game is truly stunning—more so perhaps even than with basketball. In the old game, the players' tiny wooden rackets appear to be moving in slow motion, and the ball fairly floats around the court, like a balloon. Much has been made of the difference in rackets. As a player who grew up with the old rackets and now uses the new ones, I can tell you the difference is profound. Sampras and his contemporaries, then, would destroy Drobny and his—if we could arrange such a set of fantasy matches. But what if a young Drobny, say 14 years old, were transported through time to today, and given the time to learn the modern game and use of the modern equipment? Would he have risen to the top of today's game? Drobny was by all accounts a remarkable athlete; fast, strong, mentally tough, and very competitive (he was also a superb ice hockey player). There is every reason to believe that he would still have been an exceptional tennis player had he been born 50 or 60 years later, instead of in 1921.

In sports like tennis or basketball with no quantitative standard, all we have are relative comparisons. All an athlete can do is compete against his or her contemporaries, since there is no objective measure for these sports. Chris Evert and Martina Navratilova drove each other to ever-higher levels during a rivalry that lasted nearly 20 years. Wilt Chamberlain and Bill Russell pushed each other to improve, as did Larry Bird and Magic Johnson. On the other hand, John McEnroe drove Bjorn Borg into early retirement in the early 1980s. None of these athletes had an absolute standard against which to measure him- or herself. They only had each other— relative comparisons. When we compare Drobny to Rosewall to Connors to Sampras, then, we have every reason to believe that a player like Drobny would have been able to succeed in today's vastly different game were he given the same advantages that today's players have, and a little time to assimilate them. But let's not stop there.

What about athletes in sports that do have absolute, quantitative measures? On June 26, 1954, the great British runner Jim Peters set the last of his four world records in the marathon, in a time of 2 hours 17 minutes and 39 seconds at Cheswick, England. (Just one week later, as noted above, Jaroslav Drobny would defeat Ken Rosewall at Wimbledon.) The world record in the marathon in 2003 is 2 hours 5 minutes and 38 seconds, or 12 minutes and 1 second faster than Peters's mark (about 11%). You don't have to be a rocket scientist, or a sports fanatic for that matter, to divine what would happen if the Jim Peters of 1954 went up against the 2003 record holder, Khalid Khannouchi, whose record is about 27 seconds per mile faster than that of Peters. The latter's 1954 record is in fact 20 seconds slower than the current *women's* marathon record, set by another Briton, Paula Radcliffe, in 2002. But what if Jim Peters were, as in our earlier thought experiment with Jaroslav Drobny, magically transported to the present time in his youth and allowed to train with modern runners using modern techniques and modern equipment, eat a modern diet, and so on? Would he be able to run with the fastest men, or would he be better off trying to

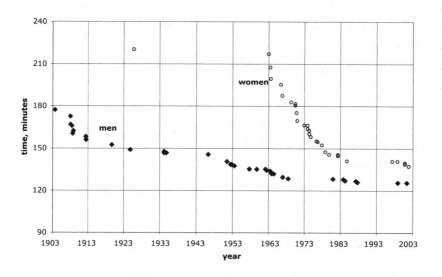

Fig. 5.3. Men's and women's marathon world records

keep up with his compatriot Paula Radcliffe? Just as I believe Drobny could play tennis with today's best men, I believe Peters would rise to or near the level of today's best male marathoners.

Take a look at the graph in Figure 5.3 of marathon world records for both men and women. In this figure, several differences can be seen between the record progressions of male and female marathoners. First, world records for male marathoners have been kept since 1904, while the first women's record was recorded in 1926. This record was not broken until 1963, and then only by a few minutes, after which the women's record began to improve at a furious pace. Between December 1963 and April 1983—less than 20 years—the women's record decreased by almost an hour and thirteen minutes! This represents a greater than 33% reduction in the record time. (Since 1983, the rate of improvement in the women's record has closely resembled the men's.) Over roughly the same twenty-year period, from 1963 to 1983, the men's record decreased by less than six and a half minutes, or just under 5%. This clearly means that women marathoners were improving faster than the men during this period—that's what the numbers say. But why? Well, the marathon was an unusual event for women prior to 1963. Once women did embrace the marathon in the 1960s, they began to train seriously. They were thus able to take advantage all at once of the techniques that male marathoners had been steadily developing throughout the twentieth century. The combination of a previously underutilized pool of talented women suddenly taking advantage of 60 years worth of training advances led to spectacular progress, and very quickly.

It is my contention that the women marathoners of the 1960s and 1970s were a little bit like the Jaroslav Drobny of my earlier thought experiment. Were the pioneering women who set the first records in the early 1960s really 33% worse athletes than those setting the records twenty years later? Of course not! They were just making up for lost time—about 60 years of it—as fast as they could, just as Drobny, or Jim Peters in the men's marathon, would do if they could magically be transported to the present while still in their athletic prime.

In Tables 5.1 and 5.2, some data from the record progressions in track and field for both men and women are recorded. Four events are considered: the 100m, 1,500m, and marathon running events, and the pole vault. For the men, progress is steady but unspectacular in the running events. Improvement in the world records in these events is greatest in the marathon, at only 0.30% per year. The pole vault is included because (like tennis) it has benefited from spectacular improvements in equipment: the pole. Since 1912, men's pole vault records have improved by 0.64% per year—or more than twice as fast as the marathon. The women's records (Table 5.2) have in every case improved much faster than the men's. In these four events, the women's records have improved roughly twice as fast, per year, as have the men's.

Table 5.1
Men's record progressions

Event	Early record	Recent record	Improvement	Improvement per year
100m run	1920: 10.6 sec	2002: 9.78 sec	14%	0.09%
1500m run	1912: 236.8 sec	1998: 206 sec	13%	0.15%
Marathon	1904: 177.3 min	2002: 125.63 min	29%	0.30%
Pole vault	1912: 4.02 meters	1994: 6.14 meters	53%	0.64%

This whole "record progression" discussion can even be expanded to include a discussion of IQ. Although it is more difficult to quantify the progression of IQ results (since the test questions and style have changed over the years), it is generally believed that IQ scores for a given population (say the British) have been increasing by more than five IQ points per decade (see Table 5.3). This trend is confirmed for many different populations and over up to six decades of results. The trend is known as the "Flynn Effect," after the New Zealand–based researcher and political science professor James R. Flynn, who first discovered and wrote about the phenomenon. It just so happens that the rate of improvement per year in IQ scores as documented by Flynn is not all that different from the rate of improvement in various track and field records as presented in the tables above—particularly for the women. These sports records are much less subject to interpretation than IQ results. While IQ test questions have changed over the years, the 100-meter dash is still the 100-meter dash. When one considers the difference in sports record progression between men and women (as shown graphically for the marathon and 1,500-meter run in the figures above), things get even more interesting. What would happen if one could take a population that had never been IQ tested and begin to regularly test their IQ? Would their "record progression" more closely resemble that of the men's marathon or the women's? My money is on the women's—I suspect we would see a rapid rate of progression.

Top runners, swimmers, javelin throwers and weight lifters all perform

Table 5.2

Women's record progressions

Event	Early record	Recent record	Improvement	Improvement per year
100m run	1924: 11.7 sec	1988: 10.5 sec	10.3%	0.16%
1500m run	1967: 257.3 sec	1993: 230.5 sec	10.4%	0.40%
Marathon	1926: 220.4 min	2002: 137.3 min	38%	0.50%
Marathon*	1963: 217.1 min	2002: 137.3 min	37%	0.94%
Pole vault	1983: 3.59 meters	2001: 4.81 meters	34%	1.9%

*Neglecting the first women's record set in 1926.

Table 5.3

IQ progression

Group	Early average	Recent average	Improvement	Improvement per year
Britain	1940: 73 IQ	1992: 100 IQ	37%	0.72%
Netherlands	1952: 79 IQ	1982: 100 IQ	27%	0.88%

measurably and remarkably better today than their predecessors. Athletes in sports without such quantitative measures of excellence, like tennis and basketball, are also clearly much better today. This is getting a little far afield from the sports topic of this chapter, but what about comparing "performers" in areas other than sports from different eras? Are Toni Morrison and Salman Rushdie better novelists than, say, Herman Melville or Jane Austen? Are Leonardo DiCaprio and Julia Roberts better actors than Bette Davis or Cary Grant? Is Stephen Hawking a measurably better physicist than Albert Einstein was?

A scientist like Stephen Hawking builds his theories on those of the scientists who came before him. Hawking didn't have to invent the theory of relativity—Einstein had already done so. Hawking is thus free to build on that theory, or to tear it down if it is wrong, or he can concern himself with other matters. Likewise, an athlete builds his or her performances on those of earlier performers. When I was a child, my friends and I all imitated Mickey Mantle or Bob Gibson on the baseball diamond, and Rod Laver or Arthur Ashe on the tennis court. The best athletes of my generation were simply those of us who did the best job imitating, and building on, the performances of our heroes. Musicians do the same thing, and so do actors and writers.

Let's assume, just for argument's sake, that Albert Einstein really was "the greatest physicist of all time," a title for which he would no doubt gain many votes from experts and amateurs alike. By all accounts, Einstein did his greatest work between 1905 and 1935. It is extremely unlikely, by the measures of greatness that we use for sports, that "the greatest" in any sport, in any sort of athletic competition, could have come from a similar era. Why is this? Why can the best physicist come from any era, while the best athlete in any particular sporting endeavor is almost certainly of a much more modern vintage?

It could be that it is simply too hard, or even impossible, to measure "the best physicist" and thus that question must be left open to argument and persuasion—there is no objective standard. There is a related argument that can also be made. In sports, particularly those with objective measures like time, the best athletes simply aim for those targets. If I am a great sprinter, and my best time in the 100 meters is only 0.1 seconds (about 1%) off the world record, my task is simple (but certainly not easy): just find a way to shave 0.1 seconds off my mark, and I'll be the greatest sprinter who ever lived. Perhaps a minor adjustment in my starting technique, or a little more leg strength (all of which can be measured), and I'll be there. There is nothing like this, of course, for physics, or writing, or painting.

Fortunately, we don't have to concern ourselves too much with who is better, Hawking or Einstein, Morrison or Melville. (Notwithstanding the "Top 10 or 100 _____ of all time" lists that those with nothing better to do are always making.) We can admire all of these people without worrying about measuring their performances. Why is this so much more difficult to do with athletes?

CHAPTER 6

MEASURING

THE

MIND

Intelligence, Biology, and Education

He's a genius with the IQ of a moron.
—Gore Vidal on Andy Warhol

Don't pay any attention to what they write about you.
Just measure it in inches.
—Andy Warhol

In June 2002, the U.S. Supreme Court ruled in a 6–3 decision that the mentally retarded may not receive the death penalty for any crime that they might commit, no matter how heinous. Mental retardation is defined differently by the various states and by the federal government, but among the most common criteria is a threshold IQ of 70. New Mexico, for example, holds that, "an IQ of 70 or below on a reliably administered IQ test shall be presumptive evidence of mental retardation." We are thus faced in the United States with a situation where perhaps the most controversial measurement ever devised, the intelligence quotient, has literally become the difference between life and death in some cases.

The Faith Quotient

Imagine a society in which all the citizens are members of a single religious faith, and that all the institutions of the society itself are controlled by the leaders of that religion. Those who possess a deep religious faith are the most revered members of the society. Young children with the strongest faith, as measured by special tests, are given accelerated religious training—these will be the society's leaders. Those children whose test results show average levels of faith will become the assistants to and the followers of the religious leaders that have just been culled from their ranks. The rest of the children, who show a profound lack of religious faith, are placed in remedial schools and relegated, eventually, to unimportant jobs.

Few of us would want to live in a society like this. Even those of deep religious convictions would find something unsettling about measuring something as personal, intangible, and elusive as religious faith—and in placing so much importance on the results of those measurements. Measuring religious faith in young children is an even more troubling notion. The idea that religious faith must be nurtured in children, and that it will grow over time, is deep-seated. It is anathema to think that each of us is born with a certain amount of religious faith, that that faith will not change over time, and that it can be accurately measured in the very young. I am, I believe, only slightly overstating the case to say that this is exactly what we have in American society today, except that instead of carefully measuring religious faith, we measure something called "intelligence."

What is different about intelligence as compared to other concepts such as religious faith, love, courage, or creativity? The difference is clear: we live in a society that places a great deal more importance on intelligence than it does on any of these other concepts. American society is, to some degree at least, a meritocracy. In a meritocracy, people are given opportunities based on their abilities and accomplishments. In a nonmeritocratic society, there isn't much need to measure intelligence. In an aristocracy, for example, people are given opportunities based on heredity. It only matters who your parents are, not whether you are intelligent (or athletic or artistic). But the ability most prized in our meritocracy is what is called "intelligence." We are wild to find out who has intelligence and in what quantities, because we are convinced that it is the best thing going.

Where to Begin?

When you read the literature related to measurement in all its manifest forms, one type of measurement turns up, again and again, often to be cast in the most sinister of roles. The measurement of intelligence, especially today, epitomizes for many all that is dark and evil about measurement. There is certainly a positive side to the intelligence measurement story, and it is possible, I believe, to consider it dispassionately. As E. L. Thorndike noted, everything that exists can be measured. But what if intelligence doesn't exist? As it turns out, when it comes to intelligence (whether or not it exists, and whatever it is if it does), it's not so much whether we measure it, but what we do with the numbers afterwards.

The measurement of intelligence has been praised as "one of the greatest scientific advances of the entire 20th century." During that same century, however, IQ and intelligence measurement were linked to both eugenics and forced sterilization. On the other hand, a great many children with learning disabilities have benefited from remedial schooling prescribed as a result of the early detection of low intelligence. Faced with such a range of outcomes, it is clear that we need a few ground rules in discussing IQ. First, it is entirely reasonable to expect humankind to try to measure something like intelligence, mental ability, or whatever you might want to call it. Next, let us acknowledge that is not surprising that intelligence would prove difficult to measure in a satisfactory manner, nor is it surprising that such measurements provoke strong emotional reactions among some of the measured. Finally, and above all, let us agree that it is not our attempts at the measurement of intelligence itself that lead to all the heartache—it's how those measurements are *used*. Measuring intelligence and then giving people carte blanche to do what they will with the results is a little bit like awarding the concessions contract at a substance abuse clinic to the local liquor store.

The Great Satan? IQ and the Measurement of Intelligence

In 1989, the IQ test was listed along with the discovery of DNA, the transistor, and heavier-than-air flight as one of the 20 greatest scientific advances of the twentieth century by the American Academy for the Advancement of Science. Now, there was no lack of outstanding scientific achievements in the twentieth century, so we can assume that the IQ test was not added to this list casually. How did the IQ test come to be held in such high esteem?

The idea of "intelligence," or the ability to learn facts and skills and apply them, arose originally from everyday observation. Just as the concept of temperature arose from observations of "hot" and "cold" (e.g., fire and ice) so the concept of intelligence arose from observations of the ease or difficulty with which various individuals were able to solve problems or master new skills such as foreign languages or mathematics. Some people seemed to be able to solve problems and master such skills with relative ease, while others struggled mightily with the same tasks. If we look at intelligence in that way, why shouldn't we try to measure those abilities? No one would argue that measuring temperature is somehow wrong or contrary to nature, so why shouldn't we attempt to measure these problem-solving skills, which we observe in nature just as clearly and as innocently as we observe differences in temperature?

Plato and the Philosophy of Intelligence

Plato, no mean observer of the human animal, classified the major aspects of the mind or soul into three parts, which he called intellect, emotion, and will. In the *Phaedrus*, he provides a vivid analogy. The intellect is a charioteer who holds the reins, while the horses that pull the chariot represent emotion and will. Intellect guides and directs, while emotion and will supply the power.

Plato also addressed the timeless question of nature versus nurture in intelligence. That he attributed differences in intellect to genetic causes—that is, nature—is made clear in his fable of the metals in *The Republic*, where he says that God created us all from different metals; gold for those fit to be rulers, silver for those who do the rulers' bidding, and a mixture of iron and brass for the working class.

Thus Plato groups us all into three categories, based on our intelligence, the content of our characters, or some combination. He is not measuring intelligence, but he is at least classifying it, and thus suggesting that it can be measured. Classification schemes like this often predate numerical measurement systems—in fact the presence of a classification system for something is an excellent indication that a measurement scale is probably not far behind. The hardness of a material, for example, existed only as a formal classification system (diamond is harder than glass is harder than steel is harder than brass . . .) until early in the 1900s, when the first of various numerical measurements for hardness was developed. Today, they are commonplace.

The example of hardness is precisely the one Lord Kelvin brings up in his lecture after his remarks about measurement and "meager and unsatisfactory knowledge." He bemoans the fact that there is no quantitative scale for hardness, even though he knows that diamond is harder than glass, and steel harder than brass (and that he can thus classify hardness). This is Kelvin wearing his engineer's hat. When scales were developed to measure hardness, they gave the engineer an important set of tools to use in designing machines that had not existed before. The modern hardness test generally involves pushing a small, hard object (often a diamond) into a material and measuring the material's resistance to being penetrated. Today, the hardness of a material is very often specified on mechanical drawings, along with other equally measurable factors such as strength, surface finish, and so on.

It should come as no surprise that Plato's initial classifications of intelligence would eventually lead to measurement scales. But in addition to classifying intelligence, and perhaps also character, Plato also implied these characteristics are innate; an iron or bronze person can never be changed into a silver or gold one. In one passage, he skates even further out onto thin ice: "Hence the god commands the rulers . . . to keep over nothing so careful a watch as the children, seeing which of these metals is mixed in their souls. And, if a child of theirs should be born with an admixture of bronze or iron . . . (they) shall assign the proper value to its nature and thrust it out among the craftsmen and farmers . . . believing that there is an oracle that the city will be destroyed when an iron or bronze man is its guardian." Plato recounts this fable as a conversation between Socrates and Glaucon in book 3 of *The Republic*. Socrates ends the fable by asking Glaucon, "So, have you some device for persuading them of this tale?" Glaucon replies, "None at all for these men themselves; however for their sons and their successors and the rest of the human beings who come afterwards."

The same tale, in increasingly sophisticated versions, has been told to succeeding generations. In *The Mismeasure of Man*, Stephen Jay Gould relates the means by which those in the West have justified the practice of attempting to rank individuals (and groups) according to their "inborn worth." He writes, "Plato relied upon dialectic, the Church upon dogma. For the past two centuries, scientific claims have become the primary agent for validating Plato's myth."

Aristotle and others have contrasted the observed behavior of an individual with the underlying capacity on which that behavior depended. Intelligence, as one of those underlying capacities, is an ability that may or may not always be observed in everyday life. This underlying capacity must be deduced from observed behaviors. We not only need to select the right behaviors to observe but also ensure that the underlying capacities are properly deduced. Measurement, when it cannot be carried out directly, is often accomplished through a process of deduction.

Is Intelligence Different?

In *The Structure and Measurement of Intelligence*, Hans J. Eysenck compares the measurement of intelligence with that of more traditional quantities in the physical sciences, such as gravity and temperature. He scolds critics of IQ who claim that "attempts to measure intelligence are doomed to failure, and even slightly absurd; that science does not deal with intangibles, like mental qualities." These criticisms are made, he claims, by those who are unaware of how measurement is done in the so-called hard sciences. If there are defects in the scales employed by psychologists, so are there in those employed by physicists. If the IQ test and measurement scale seem somewhat arbitrary, there are similarly arbitrary measuring devices and scales related to gravity, temperature, and other physical concepts. "If the measurement of temperature is scientific," he claims, "then so is that of intelligence."

It may simply be that "intelligence" is that much more difficult to measure than temperature. Or, perhaps, our inability to measure temperature exactly is far less problematic than our inability to measure intelligence precisely. Think of how we use temperature measurements in our everyday lives: the weather, the temperature inside the oven when we cook, our body temperature when we have a fever, and so on. The precision we need is easily obtained, the required instrumentation plentiful and inexpensive, the consequences of small errors of little importance (except when it comes to baked Alaska, soufflés, or global warming). Even in a high-tech manufacturing plant, modern temperature measurement technologies for furnaces, ovens, cryogenic storage containers, and so on, are more than adequate.

Putting aside the difficulties in its measurement, we can define something called "intelligence" and contrive ways to measure it, but the practical applications of those measurements are much more difficult than those generally associated with temperature (global warming notwithstanding), or gravity, or a whole host of other concepts. We use intelligence measurements in order to classify ourselves. Not necessarily to separate, in Plato's terms, the rulers from the rest of us, but at least (in American society today) to decide who will gain admission to the best schools and thus have the best chance for success in our society. In other words, we use our ability to measure certain aspects of our *nature* (intelligence) to decide, to some degree at least, how we will be *nurtured* (schooled). Crucial decisions, indeed. No wonder that such measurements are so controversial. If intelligence testing were just being done for our information, much as a career interest survey is done—there would be no problem. But when you tell parents, "Your child fared poorly on the intelligence test and is clearly not very bright, so we're going to put him in the class with the slow learners," it's another matter entirely.

Eysenck compares the measurement of intelligence with that of gravity. He notes that we can define the concept of gravity by

1. referring to the actual phenomena that the concept seeks to explain and predict, namely, falling bodies;
2. devising theoretical explanations such as Newtonian gravitation, the graviton (a theoretical particle in quantum physics supposed to carry gravitational force), or warping of space-time; or
3. employing a formula that tells us how to measure the force involved, that is, define the concept in terms of its measurement.

We can define intelligence, too, in similar ways, by

1. pointing to the actual ways in which people manifest their intelligence, for example, by solving problems;

2. devising an entirely theoretical explanation of the concept of intelligence; or

3. defining the concept in terms of what intelligence tests measure.

Why, Eysenck asks, does it seem ridiculous to some of us to define intelligence in these ways, but not gravity?

A crucial problem in the development of the concept of intelligence as a scientific theory, and indeed in the measurement of intelligence, is the need to distinguish between mental ability (intelligence) and knowledge. It is assumed that we can only measure performance, and not the underlying mental ability, which must thus be inferred from the performance. In 1928, E. L. Thorndike noted that all the measurements of intelligence then available were measures of some product produced by the person in question, "A is rated as more intelligent than B because he produces a better product, essay written, answer found, choice made, completion supplied or the like."

In the same way, we cannot directly measure the mass of the Earth but must infer it from the Earth's "performance" (in particular as it relates to the gravitational forces the Earth exerts). We can deduce the mass of the Earth without measuring it directly because we have Newton's theories at our disposal. Deducing intelligence from observed performance is more difficult—there is no universally accepted theory of intelligence analogous to Newton's theories of motion. (The Earth weighs about 6×10^{24} kilograms, by the way.)

One Intelligence or Many?

In psychometrical circles, there is an old debate about whether a person possesses one overall "intelligence" or many individual intelligences. Those who favor the multiple-intelligence approach have produced many such lists. The American psychologist and psychometrician L. L. Thurstone's list is as well known as any. Thurstone theorized that there are seven primary mental abilities: verbal comprehension, word fluency, number facility, spatial visualization, associative memory, perceptual speed, and reasoning. This list was the result of careful statistical analysis of intelligence test data—that is, of tests intended to provide one single measure of intelligence. Thurstone was thus among the first to argue that there are multiple ways in which someone can be intelligent—and that they can all be measured.

Statistical analyses of intelligence test data (such as the work of Thurstone) have been performed for a long time. Given the same set of data, one can make a convincing, statistically sound argument for a single, overriding intelligence (sometimes called the g factor) or an equally sound argument for multiple intelligences. In *Frames of Mind*, Howard Gardner argues that "when it comes to the interpretation of intelligence testing, we are faced with an issue of taste or preference rather than one on which scientific closure is likely to be reached." The question of taste or preference is an important one, I believe. As discussed in Chapter 3, we have (particularly in America) a fondness for overall measurements and rankings. That is exactly what a single intelligence measurement offers.

Intelligence and Empiricism

If someone made a movie about Sir Francis Galton (1822–1911), it might be called *Charles Darwin's Smarter Cousin*. Much less celebrated today than Darwin, Galton is best known for his studies in human intelligence and for advances in the field of statistics. His wealth allowed him the freedom to pursue his scientific interests in a multitude of disciplines, which, combined with his brilliance and determination, resulted in important advances in many of those fields, including significant contributions in meteorology and in the use of fingerprints to identify people. Galton, unfortunately, also invented the term "eugenics" and was an early, enthusiastic advocate of the regulation of marriage and family size based on the parents' hereditary gifts.

Galton was above all a measurer. Stephen Jay Gould called him an "apostle of quantification." In that regard, Galton appears to have been in total agreement with his contemporary Lord Kelvin. Galton believed that anything could be measured, and he even published a scholarly paper entitled "Statistical Inquiries into the Efficacy of Prayer." (Similar inquiries continue. In 1997, at least 30 medical schools in the United States offered courses relating to religious faith and the role of prayer in healing.) Seemingly as much for his own amusement as anything else, Galton made maps of the British Isles showing the distribution of the beauty of women (he concluded the most beautiful women were in London), and he developed and employed criteria for quantifying boredom (as measured by the frequency of fidgeting), which he used to rate presentations at the Royal Geographical Society.

Through his measurements, classifications, and statistical analyses, Galton developed rankings of individuals based on their physical and mental capacities. He developed links between genealogy and professional accomplishment and became a strong believer in the heritability of almost everything. Having read Galton's *Hereditary Genius*, Charles Darwin wrote to his cousin, "You have made a convert of an opponent in one sense, for I have always maintained that, excepting fools, men did not differ much in intellect, only in zeal and hard work." Galton's reply was: "The rejoinder that might be made to his remark about hard work, is that character, including the aptitude for work, is heritable like every other faculty."

Francis Galton's work represented a shift in the investigation of intelligence, which until that time had been closely allied with philosophical inquiry. To Galton, intelligence was just another thing that he could measure. (Someone unafraid to measure the efficacy of prayer seems unlikely to have been deterred by the concept of intelligence.)

Galton's work on intelligence, however, shares at least one fault with the work of later investigators such as Alfred Binet, the father of IQ: it was, as Gardner says, "blindly empirical." That is, it was not based on any theory of how the mind might or might not work. There is a complex relationship between theory and measurement in the sciences. Thomas Kuhn, among others, grappled with this relationship. Kuhn wrote that "the route from theory or law to measurement can almost never be traveled backwards. Numbers gathered without some knowledge of the regularity to be expected almost never speak for themselves. Almost certainly they remain just numbers." Kuhn admits that there are counterexamples—he cites Boyle's Law relating gas pressure and gas volume, and Hooke's Law relating the displacement of a spring to the force applied to it. These important physical laws are both the direct result of measurement (the numbers came first, and then the theory followed). But Kuhn claims that it is far more typical for measurement to follow a qualitative theory, as examples of which he cites Einstein's theory of relativity and Newton's three laws of motion. Concerning Newton's laws, formulated in the late seventeenth century, Kuhn says that for the better part of 100 years afterward, a great deal of superb science was devoted just to the experimental confirmation of those famous laws.

Intelligence measurers, in Kuhn's way of thinking, always seem to have been doing things backwards. Lacking a reasonable theory of how the mind works, they simply measure, and measure, and measure some more. In and of itself, this argument goes, such measurements are unlikely to lead to sound theories of how the mind works and, as a consequence, are very likely to result in misinterpretation of the numerical results. Howard Gardner notes that "the I.Q. movement is blindly empirical. It is based simply on tests with some predictive power about success in school and, only marginally, on a theory of how the mind works." The measurements themselves aren't intrinsically troublesome. How they are applied (e.g., in the case of school choice or the death penalty) is another matter.

Binet, IQ, and America

Alfred Binet (1857–1911), a French psychologist, developed what became surely the most popular measure of intelligence, and one of the most recognizable overall measurements of any kind, the IQ, or intelligence quotient. When Binet's work was brought to America and adapted and expanded, he was first hailed as a hero. Later, when some of the bloom came off the IQ rose, he was vilified. In truth, what he has mainly been is misunderstood.

Binet began his work in intelligence measurement by employing the methods of his celebrated countryman Paul Broca. Binet began with Broca's conclusion that head size and intelligence were correlated (the bigger the head, the smarter the person), but his own craniometric studies did not confirm Broca's results.

Binet's subsequent attempts at measuring intelligence were thus psychological, not biological. His most famous work began when he was commissioned by the minister of public education to help develop techniques for identifying which children needed special help in the classroom (what we might refer to today as children with learning disabilities). Binet, an eminent theoretician, chose an entirely empirical approach. He developed a series of tasks of increasing difficulty for children to perform. Proceeding gradually from easy to difficult tasks, children continued performing until they reached a level of difficulty at which they could no longer succeed. That level, the level of the maximum difficulty of task the child could perform, would then be used to determine the child's "mental age." The German psychologist Wilhelm Stern suggested in 1912 that the mental age determined by Binet's tests be divided by the child's chronological age (and then multiplied by 100) to yield the intelligence quotient, or IQ. Thus, a 10-year-old child who achieved a mental age of 12 on Binet's test would have an IQ of 120.

From Binet's writings, it is clear that he believed IQ to be a measurement of limited utility, although he was justly proud that this new tool fulfilled his charge from the minister of education—it really did help identify children with learning disabilities. Binet wrote clear guidelines for the use of his tests, and provided stern warnings against their misuse. He noted that IQ scores were simply a rough guide to help identify children with learning disabilities so that they might get special help. IQ was not intended to be used for ranking normal children, nor did it measure anything innate or unalterable. Young children who scored low on the test should not be labeled as inherently incapable. These IQ scores, he found, had some utility for this narrowly defined purpose, but they did not correlate to any particular theory of intellect, and they certainly did not define anything like "intelligence."

When Binet's measurement technique and the resulting scale, the IQ, made the leap across the Atlantic, his warnings and guidelines did not. Had we in America heeded Binet, we might have been spared, as Gould wrote, "a major misuse of science in America."

The perversions that were visited upon Binet's creation in America have little to do with the measurement of IQ as Binet defined it. Reading the early history of IQ in America, it almost seems as if the early American pioneers, men such as H. H. Goddard, L. M. Terman, and R. M. Yerkes, were simply waiting for something quantitative, some sort of measurement whose results they could bend to suit their purposes. IQ gave them a powerful tool with which to promulgate their hereditarian, even racist, views.

It is amazing what has happened to a simple, pragmatic test, designed for a simple purpose: to quickly identify children in need of special education. What might have happened if Binet's test had not resulted in a numerical scale (the IQ scale)? What if, instead of IQ, the Binet test result had been a simple yes or no? Yes, this child needs special education, or No, this child should study in the normal class. If Binet's test had only classified the children, instead of ranking them, its subsequent perversion in America would have been more difficult. Having the ranking scale makes it all so much more powerful.

Binet's test was originally administered orally—a trained supervisor tested each child individually. In America, with the advent of the Stanford-Binet test (Terman's main contribution) the intelligence test could be mass-marketed to large groups by a single test administrator. Eventually, it became 100% multiple choice and could thus be machine-graded when that technology became available in the 1930s. (As of the time of writing, intelligence tests are 100% machine-graded multiple-choice exams, as are American college admission tests such as the SAT and ACT, but Binet's labor-intensive testing is not simply a quaint souvenir of a bygone era. In France, the national high school diploma exam, known as the *baccalauréat*, remains a labor-intensive hand-written affair in which the students write long essays and perform complex mathematical calculations, which require the painstaking labor of thousands of teachers every year to prepare, administer, and grade.)

Binet's test was only for children, but it was extended to adults in America. (This began in earnest with the use of IQ tests by the U.S. military, as promoted by Yerkes.) Little by little, the results of the IQ test, with their convenient numerical scale, became synonymous with "intelligence" in America. One could scarcely hope to find a clearer example of reification. Against Binet's specific instructions, American IQ tests became a means for ranking normal children.

Finally, and perhaps most important, IQ in America became not simply a means of identifying and helping children with special needs, but a means of reinforcing racial and ethnic stereotypes, of evaluating people for specific purposes (school, the military, job placement, and social companionship), and of culling the most promising students from the top of the scale while shunting aside those at the bottom.

Gradually, in fits and starts, psychologists and educators in America have started to come to grips with IQ and intelligence testing. As stern a critic as Gould himself admits (*The Mismeasure of Man* was published in 1981) that in some instances IQ testers in America have even reverted to many of Binet's original goals—Gould's own son, who suffered from learning disabilities, benefited from such testing.

The limitations on intelligence testing are better known and more accepted today. For example, it is generally recognized that intelligence tests tend to be skewed toward individuals in societies where schooling is valued and where there is an emphasis on paper-and-pencil tests. IQ thus tends to be a good predictor of success in school, but less so for success in other areas of life, particularly when social and economic background are taken into account. It also appears that Binet may have been right when he cautioned that IQ should be used only to identify children with learning disabilities, or those at the lower end of the IQ scale, and not as a means of ranking normal (or above-average) children. A variety of studies have shown that IQ is an unreliable predictor of a successful career for those whose IQ measurements are significantly above average.

At the very least, IQ shows what can happen when an overall measurement becomes too powerful. IQ is an important predictor of success in school, but it is not the only such predictor, and it is of limited utility for other purposes. To me, IQ is a little bit like the height of a basketball player. All other things being equal, a taller basketball player will have advantages over a shorter one. But all other things are rarely ever equal, and "shorter" basketball players (Michael Jordan is "only" six feet five inches tall) are often superior players. What sense would it make to rank prospective basketball players *solely* on their heights? Height is a "quick and dirty" way to get some idea about potential basketball prowess, just as IQ is a quick and dirty way to get some idea about intellectual prowess.

From all this we are left to conclude, with regard to intelligence and its measurement, that

- there is something called "intelligence," or innate intellectual ability, that exists in human beings;
- intelligence may be more than one thing, and may even be a great many things;
- some people possess more of this thing or these things than others;
- the amount of this thing or these things that each of us possesses may vary over time;
- we can measure the thing or things that make up intelligence, although how to make and interpret these measurements remains controversial—many believe we don't really know what we are measuring; and

- the results of intelligence measurement can be useful, but these results are easily abused—sometimes with horrifying consequences.

It seems to me that at least some of the difficulty with the measurement of intelligence lies in the personal nature of the measurement. When our intelligence is measured, many of us assume that, "This is all I've got, and I can't get any more, no matter what I do." We can deal with the knowledge that, for example, we'll never be any taller than our adult height. If a physician tells a mother that, based on his measurements, her young son will likely be no taller than five feet six inches when full grown, she can accept that. Lots of happy, successful men are not very tall. But to learn as a parent that your young son's innate, unchangeable "intelligence" places him in the bottom 20% of his peer group is pretty scary news indeed.

Biology and Intelligence

Throughout the story of the development of the concept of intelligence and of its measurement, there runs the thread of biology. If intelligence is something partially or wholly innate, then almost by definition, it must have biological origins. If that is true, then it would follow that intelligence could be determined, directly or indirectly, through the measurement of biological factors. There is a long history of such measurement efforts, and it continues to this day.

One of the earliest became known as phrenology. Franz Joseph Gall, as a student in the late 1700s, noted certain coincidences between the physical characteristics of his classmates and their intellectual capabilities. For example, boys with prominent eyes seemed to have good memories. Later, as a physician and scientist, Gall greatly expanded his observations and organized and codified them, which gave birth to the discipline of phrenology. Phrenology's chief tenet is that human skulls differ from one another, and therefore the brains they contain are of different sizes and shapes. Further, since different areas of the brain are in charge of different functions, it should be possible, by carefully observing the size and shape of the skull, to determine the mental strengths, weaknesses, and idiosyncrasies of individuals. Gall and his colleague Joseph Spurzheim listed thirty-seven different powers of the mind they claimed could be indirectly measured through the size and shape of the skull, including secretiveness, hope, reverence, self-esteem, language, (musical) tune, and sensitivity to visual properties like shape and color.

In the early 1800s, phrenology was wildly popular in both Europe and the United States, and not just among scientists. Even today, books on the subject are still in print, and one can order phrenological maps and models of the skull (the latter exist mainly for their nostalgic value or as gag gifts, it would seem). More recent developments in the study of the brain have even resulted in what some have called a neophrenological movement.

Looking back, it's easy to make light of some of Gall's and his colleague's claims. The overall size of the skull (and thus the brain), for example, is not necessarily correlated with intellectual power; many geniuses had small heads and brains, and people with extremely large brains are sometimes quite unintelligent. In addition, the size and shape of the skull maps inexactly to the size and configuration of the various regions of the brain it contains.

However, some of Gall's contributions were important. He was among the first modern scientists to stress the localization of various functions to various parts of the brain. Also, that we have not yet precisely defined the relationship between size, shape, and function does not mean that we shall not someday be able to do so—it is possible that Gall really was on to something in this respect. Finally, Gall was among the first to propose that general mental powers, such as perception, memory, and attention, do not exist, but rather that there are several forms of each of these for each of the various intellectual faculties, such as language, music, or vision. This viewpoint calls to mind the theory of "multiple intelligences" espoused by Howard Gardner and others.

After Gall, the next important contributor to understanding the relation between biology and intelligence was Pierre-Paul Broca (1824–80), who showed that lesions in a certain area of the left anterior portion of the human cortex caused a breakdown of linguistic capabilities—a condition known as aphasia. Broca and others went on to show that other lesions in specific locations in the left hemisphere could impair the ability to read, for example. What Gall was attempting to infer through his phrenological measurements, Broca seemed to have proven through direct observation of the brain.

Broca, like Gall, was also a phrenologist (perhaps the term "craniologist" would be more precise). One of Broca's core beliefs was that brain size and intelligence were strongly correlated. Broca was an exceedingly careful experimentalist, and he generated large volumes of outstanding data. For example, he labored for months refining his preferred technique for measuring skull volume, which involved pouring fine lead shot into the skull so as to carefully fill every nook and cranny. Broca also personally weighed hundreds of brains that he removed from cadavers he had autopsied. Many of Broca's conclusions from these studies mirror those of the early American intelligence testers. Successful white males had the largest brains, followed by women, blacks, and poor people. Gould, who studied Broca's original works extensively, concludes that Broca, in developing his conclusions, "traversed the gap between fact and conclusion by what may be the usual route—predominately in reverse." That is, although Broca's data are beyond reproach, they were gathered selectively, according to Gould, and then manipulated, perhaps unconsciously, to support his already-formed conclusions.

Recent developments in techniques such as MRI have given rise to what some have called "neophrenology." Unsurprisingly, the term has a negative connotation, since phrenology itself is discredited today. To have one's work in neuroscience or neurobiology labeled "neophrenology" today is to be dismissed at least as irrelevant or at worst as a quack. But there have been some breathtaking advances in our understanding of how various areas of the brain function in relation to various tasks such as reading, speaking, and doing math. The goal of this work is to link the increasingly sophisticated things we can measure about the brain to the more qualitative aspects of psychology and teaching. For example, functional magnetic resonance imaging (or fMRI) is a technique that measures regions of increased blood flow in the brain when those regions are being "exercised." Studies of the blood flow in the brains of dyslexics, for example, show that their brains function differently during reading than the brains of nondyslexics. Fascinating stuff, but what good is it to the dyslexic? At the very least, this sort of measurement technique could aid in the early diagnosis of learning disabilities such as dyslexia (perhaps even before the child has begun to read), thus allowing remedial treatments to begin much earlier than they would have with traditional methods of diagnosis.

The utility of such novel measurement techniques is limited only by our imaginations. Perhaps therapies will be developed that will allow us to improve the blood flow to underutilized areas of our brains, and thus make us smarter. These therapies could involve surgery, drug therapy, or maybe just "brain exercises." (The popular Brain Gym program tries to do this by promoting exercises that involve the body and both brain hemispheres in ways that are said to increase, among other things, concentration, memory, and physical coordination.) Just as an athlete can measure the effectiveness of a weight-training program in a variety of ways, we may be able to measure the efficacy of educational programs through brain-imaging techniques such as fMRI.

Passing the Test: Measurement in Education

Perhaps we cannot change our innate intelligence, any more than we can change the color of our eyes. Thank heavens, then, that we can change what we *know*. No controversy there! The process of changing what we know is called learning—which is something that all of us do everywhere, every day. Learning is so important that we as a society have devoted a vast, formal institution to it—our educational system. Today, it is probably more accurate to refer to this as the education industry. Measurement is a huge part of this industry. The ways in which students, teachers, their administrators, and indeed the schools themselves are being measured are changing every day. In some ways, the measurement revolution in schools mirrors that in business and sports described earlier.

Few fields have seen a recent explosion in measurement activity like education, especially in the United States. It seems to me that along with my students, my school, and the rest of my industry, I am simply being "measured to death." If pressed on the subject, most students in America would probably admit to feeling the same way, and so would their teachers and administrators.

Work All Problems, and Show All Your Work.
You Have One Hour.

As a student, I wrote stories, reports, and essays, I turned in homework assignments, and I gave oral presentations. All of these were graded—measured—by my teachers. But all of those measurements pale in comparison to the most important way in which all students are measured—the written test. We start taking written tests in school at such a young age, becoming so adapted to them, that I think many of us don't realize how ridiculous they really are. A written test is an artificial creature; it is unlike almost any other human endeavor. This didn't really dawn on me until I began graduate school, after having worked for many years as an engineer. As an older (today the term is "nontraditional") student, I had to make lots of difficult adjustments when I became a graduate student. For one thing, my salary and benefits practically disappeared when I returned to school. Even worse, I went from having a private office with a phone and a secretary to sharing a phoneless, computerless battered old desk flung into the corner of an engineering lab affectionately known to my fellow (and much younger) graduate students as "the Black Hole."

All that was bad enough, but one of the most difficult adjustments of going back to school was taking written tests again. As the clock ticked away while I desperately tried to solve some intractable math problem on my first test in graduate school, I remember thinking how contrived it all seemed. Only a few weeks earlier, I had been a respected engineer, making daily decisions that affected quality, safety, productivity, and profitability. How did I make those decisions? Certainly not by going into a room all by myself, cutting myself off from all sources of information except what was between my ears, and scribbling furiously on a clean sheet of paper for a rigidly defined (and always too short) period of time.

Yet that is exactly how a written test works. A written test is artificial—like a color not found in nature. I am certainly not the first person to observe that the skills that allow a student to excel at written tests are not necessarily those that will allow him to excel in a professional career—no less an authority than Albert Einstein wrote an article about final exams entitled "The Nightmare," in which he advocated their abolition. And as noted above, there are other ways to find out to what extent a student has mastered a subject. So why do we give written tests? One crucial reason is that they are so easy to administer compared to other instruments for measuring student progress. I can grade written tests (real write-out-the-answer tests—not multiple choice tests) three to five times as fast as I can grade students' written reports. That's a huge motivating factor for the typical harried teacher.

The ease with which a measurement (any measurement—not just those in education) can be made is an underappreciated factor in determining to what extent a measurement is utilized. A measurement that is too difficult to carry out is unlikely to become popular. For example, I have always wanted to try giving my undergraduate students oral examinations instead of or in addition to their written ones. I think I could really measure what they know and don't know that way (although I'm sure they would be horrified by the prospect). But giving each student an individual thirty-minute or one-hour oral exam, rather than giving the whole class a one-hour written exam simultaneously, is too daunting. So I give written tests like everyone else. It's easier that way.

Measurement in education is certainly nothing new. Students have been measured (graded) in the United States using written examinations since at least 1845. The twentieth century saw a series of revolutions in the standardized testing of students, where the same test is administered to students from many schools, and perhaps even across the country. These tests are used to assess the minimum competency of students to be promoted to the next grade level or, in several states today, to qualify for a high school diploma. Perhaps the most familiar use of these tests is as part of the process of admission to a college or university.

The stakes of measurement in the schools has changed radically in recent years, however. In "high-stakes testing" (as it has come to be known), not only the future of the students is determined, but also that of the teachers, their schools, and even the value of real estate in the school district. (Property values can rise and fall with the reputation of the schools in that district, which in turn depends on standardized test scores.)

Texas has been a leader in high-stakes testing. In 1990, the state introduced the Texas Assessment of Academic Skills (TAAS). TAAS results have consequences for everyone associated with a school: its students, teachers, administrators, and the school board members. High-stakes testing is controversial. Detractors claim the tests typically do a poor job of measuring school quality—which is very different (and probably more difficult to measure) than evaluating the individual student.

Many education experts also warn against relying on a single test to make critical decisions, such as whether a student will move to the next grade or graduate from high school. "A test score, like other sources of information, is not exact," a National Research Council report said in 1999.

"It is an estimate of the student's understanding or mastery at a particular time. Therefore, high-stakes educational decisions should not be made solely or automatically on the basis of a single test score, but should also take other relevant information into account."

In terms of admission to college and graduate school, after standardized test results, the other measurement of most relevance to the student is the grade point average, or GPA. The GPA is not standardized (in spite of the fact that a majority of schools use the same four-point scale) and it has come under strong criticism from some. David Brooks of the *Weekly Standard* believes that the grade point average is "one of the most destructive forces in American life today." He believes its consequences are far more injurious than another favorite whipping boy, the Scholastic Aptitude Test (SAT). This is because top universities and graduate schools require nearly perfect grades for admission, and to achieve straight A's, students would do well to avoid developing a passion for any particular subject, and instead focus on mastering the method of simply making A's in all their classes.

If Brooks is right about this, then what we really have is a measurement problem, pure and simple. We're either measuring the wrong things in our students, or we're not doing a good job measuring the right things. This doesn't mean there are easy solutions to the problem Brooks describes. We tend to measure the things that are easy to measure in ways that are easy to administer. After all, throughout its history, the SAT has remained a 100% multiple-choice test (although a short written essay will be added for the first time in 2005).

Aptitude, Assessment, and Higher Education

Measuring knowledge is not without its problems and controversies—far from it. But it suffers from one less problem at least than intelligence measuring: knowledge can be directly measured. Intelligence, as Eysenck noted, must be inferred from measurements of knowledge.

The problem is that many of the measurements in our educational system do not pretend only to measure knowledge, but also—or instead— intelligence or aptitude. Our ability to perform well on measurements of knowledge (such as the GPA) and aptitude (such as the SAT) is vitally important to our success in society. In his 1999 book, *The Big Test: The Secret History of the American Meritocracy*, Nicholas Lemann describes "a thick bright line [that] runs through the country, with those who have been to college on one side and those who haven't on the other." One's position on one side of that line or the other is a more reliable predictor of things like income and attitudes than almost any other line that might be drawn, including those based on region, race, age, sex, or religion. As a consequence, higher education becomes a huge focus for parents and their children, and "a test of one narrow quality, the ability to perform well in school, stands firmly athwart the path to success."

The measurements of knowledge and intelligence associated with higher education in America are deadly serious business. "The ability to perform well in school," in Lemann's terms, means the ability to succeed or perform well at the most important of the many measurements we employ to evaluate our students.

In *The Big Test*, Lemann tells the story of the transformation in American education that began near the end of World War II, and that has, in his view, resulted in the "thick bright line" that divides American society. He contrasts the modern state of affairs, described above, with the vision of American society at the end of World War II of men like Henry Chauncey, the first head of the Educational Testing Service, or ETS, which still administers the SAT today. Chauncey and his colleagues believed that the essence of American greatness was the same social equality that Alexis de Tocqueville had found so remarkable early in the nineteenth century—the kind of social equality that was unimaginable in de Tocqueville's France or in any other country. The United States had no rigid class system in de Tocqueville's time, and it was thus in better position to take full advantage of every American's talents.

But Chauncey and others found that American society had taken a serious turn for the worse during the early twentieth century. The western frontier of the country had finally been reached, it was becoming increasingly industrial, the cities were crowded with immigrants, and socialism was on the rise. Chauncey and his colleagues believed that the idea that individual opportunity was the highest good was on the wane. What was worse, a distinct American upper class was becoming stronger and stronger. The "career path" for this upper class began at prestigious New England boarding schools (where the children of the privileged could be enrolled at birth) and continued at Harvard and the other Ivy League colleges, where rich young men, their servants in tow, went to party and play sports at the schools their fathers had attended. These privileged young men then went on to comfortable careers at venerable institutions such as Wall Street investment houses, the most prestigious law firms, the Foreign Service, research hospitals, and university faculties. This all-male, Eastern, Protestant, privately educated upper class held a stranglehold on such institutions. It tended to be snobbish and prejudiced in the extreme and valued a vague, difficult to measure quality called "character" above all else, while tending to ignore intelligence and scientific expertise.

Intelligence and scientific knowledge were just what Chauncey and his colleagues believed were the most important traits in postwar America. How to effect a change and dethrone the elite upper class? Lemann argues that Chauncey, his colleagues, and the ETS were able to accomplish this seemingly impossible goal by transforming the American educational system—in large part by changing the way knowledge and intelligence were measured and the ways in which these measurements were used.

The rise of the SAT is linked to a program begun at Harvard to try and break the vicious cycle of the elite upper class described above. Harvard's president, James Bryant Conant, wanted to establish a scholarship program based solely on academic promise, and not on wealth, heredity, or educational pedigree. Today, this hardly seems like a revolutionary idea, but in 1933, it definitely was. In order to put his program into place, Conant needed some means of measuring the academic promise of scholarship applicants, and that was where the SAT came in. Conant was keen to measure scholastic aptitude (the ability to succeed in school) and not scholastic achievement (the knowledge already gained by a student). The test Conant chose was the SAT, which in those days stood for "Scholastic Aptitude Test." (That changed; it later became the "Scholastic Assessment Test," and today the acronym has no formal expansion at all: SAT simply stands for . . . SAT.) It is hard to fault Conant's motives in choosing a test that purported to measure aptitude over achievement. After all, he wanted to award his Harvard National Scholarships, or "Conant Prizes," to those students with the best chance to succeed academically at Harvard, even if they came from underprivileged backgrounds (e.g., poor schools, lack of nurturing families, and so on) and thus showed meager academic achievements up to that point. He was, after all, beginning this new program during the height of the Great Depression. Underprivileged students were not hard to come by.

The SAT, the test Conant chose to be his primary measure for the new scholarships, was developed by Carl Brigham, an ardent eugenicist (who later in his career eloquently recanted his eugenicist views) and colleague of some of the early American IQ promoters, such as Terman and Yerkes. In order to understand and combat what Brigham perceived as the accelerating decline in American intelligence, he adapted a version of the IQ test then being used by the U.S. Army for testing students at several universities, including Princeton. The main difference between Brigham's test for university students and the Army's IQ test is that Brigham's test was more difficult. It was still essentially an IQ test—a test of aptitude and not achievement. Thus, in 1926, the SAT was born from the Army IQ test.

The similarities between SAT and IQ in those days were striking. The SAT result was a single overall score, and Brigham even published a scale for converting SAT into IQ. Today, as most of us are well aware, the SAT results in two scores, one for verbal ability and one for quantitative or mathematical ability. (A third score, for writing, was added in 2005.)

Conant's adoption of the SAT for his new scholarship program was one important factor in its rise in popularity. Another was the development of machine grading in the 1930s. Early versions of the SAT were mostly multiple choice in the 1920s, but they still had to be graded by hand. The by now all-too-familiar score sheet in which one fills in circles corresponding to multiple-choice answers with a no. 2 pencil for machine grading was developed by Reynold B. Johnson, who took his invention to IBM, which refined and marketed it.

Between the adoption of the SAT by Conant at Harvard in 1933 and in the years just after World War II, there were many interesting developments, and much behind-the-scenes maneuvering took place in the education testing field. This had several important results. One of these was that the SAT, which remained essentially an IQ test, became the testing standard for American higher education. Another result was that the SAT was developed and administered by Conant's old Harvard buddy Henry Chauncey, through the ETS, which Chauncey ran in Princeton, New Jersey. Chauncey was an inveterate measurer much in the spirit of Galton. His

110

measurements of choice were tests. He believed fervently in the power of testing to transform American education and thus American society.

Conant's dream, begun with the Harvard National Scholarships, had largely come true. He had helped establish a system of "opportunity for all" in higher education—anyone who did well on the SAT would have the chance to go to college. (When the ETS was established in 1948, only about 6% of Americans over the age of 25 held a college degree. The figure was 24% by 1998.) At the same time, the popularity of the SAT further reified "intelligence" and allowed for its ranking, thus providing the numerical means, not only for selecting the best students for the best training, but also for shunting aside those who did not fare as well. As Lemann puts it, "This was the fundamental clash: between the promise of more opportunity and the reality that, from a point early in the lives of most people, opportunity would be limited."

On June 23, 1926, the date of the first administration of the SAT, about 8,000 students took the test. Today, over a million students a year take the modern SAT. The College Board, for which ETS administers the test, says that today's SAT "assesses student reasoning based on knowledge and skills developed by the student in school coursework." ETS has thus abandoned all pretense that the SAT measures the intrinsic aptitude of students or any other IQ-like qualities. Beneath the surface, however, SAT tries to do what it has always done; reliably predict the grades of first-year college students. Improving the "validity" (to use the tester's term of art) of the SAT—its ability to predict those first-year college grades—is and always has been one of the most important tasks at ETS. Validity, it should come as no surprise to learn, is measurable. It turns out that SAT scores, when combined with high school GPA, have high validity—this combination is a good predictor of success in the first year of college. SAT scores by themselves have less validity.

The SAT is in the midst of a major change. The "new SAT" has added a third score, for writing, to go along with the familiar verbal and math scores, something many observers believe was prompted by its threatened abandonment by the University of California (ETS's largest customer). Unsurprisingly, the College Board makes no mention of its difficulties with California, but says simply: "The new SAT will improve the alignment of the test with current curriculum and institutional practices in high school and college. By including a third measure of skills, writing, the new SAT will reinforce the importance of writing throughout a student's education and will help colleges make better admissions and placement decisions."

Most of the attention focused on the new SAT has centered on the writing portion, and within that portion of the test, on the written essay. This will be the first part of the SAT that cannot (at least not yet) be graded by a machine. Instead, each student's response to the 25-minute essay test will be electronically distributed to graders trained for the task. ETS has assembled an armada of thousands of such testers, many of them high school teachers, throughout the country. Each essay will be graded, in a period of one to two minutes, on a one-to-six scale. Steps will be taken to eliminate bias in individual graders. For example, each grader will receive, hidden among the student essays to be graded, several standardized essays for which ETS has already determined the "correct" grade. If an individual grader's marks for such standard essays are consistently below or above those of ETS, that grader's marks on the actual student essays will be adjusted up or down accordingly.

No Measurement Left Behind

On January 8, 2002, President Bush signed the No Child Left Behind Act, which includes a set of sweeping reforms of elementary and secondary education in America. Measurement is the linchpin of No Child Left Behind. The law says that we "must measure every public school student's progress in reading and math in each of grades 3 through 8 and at least once during grades 10 through 12." Required measurement of progress in science will be phased in beginning in 2007. In addition, the law requires school districts to provide parents with "report cards" on schools, "telling them which ones are succeeding and why." The schools' report cards will include "student achievement data broken out by race, ethnicity, gender, English language proficiency, migrant status, disability status and low-income status."

No Child Left Behind is controversial for a variety of reasons, but what strikes me is the emphasis on measurement. The United States, uniquely among the industrialized countries of the world, has a highly decentralized system of education. In our system, a great deal of control rests with the 15,000 or so local school districts. This local control is jealously guarded and is so entrenched that it is regarded as a part of the American Way. With No Child Left Behind, President Bush and his education advisors are trying to blend a highly comprehensive, nationally mandated measurement program with a system based on local control. It will be interesting, to say the least, to see how this plays out.

The extent to which measurement permeates all aspects of education, as exemplified by No Child Left Behind, continues to increase, much as it does in the worlds of business, sports, and elsewhere. The stakes attached to these measurements increase apace. Standardized tests once affected only the individual student. Today, standardized testing has repercussions throughout society, even extending, as noted earlier, to neighborhood real estate values.

I find all of this a bit bewildering. Like most teachers (believe it or not), I want to help students. It does me no good to throw up my hands and conclude, for example, that all intelligence testing is evil, or that standardized testing for school, whether it purports to measure knowledge, intelligence, or both, is impossibly unfair, or that the emphasis on school grades and the GPA is foolish and dangerous.

My students—and I myself, my colleagues, and my school as well—are going to continue to be measured. If I want to help my students, I have to help them succeed at the most important of these measurements. At the same time, I hope to instill in my students a love of knowledge, an open, informed curiosity about the world, and a zeal for lifelong learning. We don't measure those things, however. At least not yet.

CHAPTER 7

MAN

The Measure of All Things

*To define genes by the diseases they cause is about as absurd
as defining organs of the body by the diseases they get. . . .
It is a measure, not of our knowledge, but of our ignorance.*
—Matt Ridley, *Genome: The Autobiography of a Species in 23 Chapters*

When Richard Buckland confessed to the rape and murder of 15-year-old
Dawn Ashford, the police in Leicestershire, England, thought the case was
closed. In addition to his confession, Buckland knew details about this 1986
crime that had not been released to the public. But Buckland refused to
confess to the rape and murder of another 15-year-old girl, and the police
were convinced the same man had committed both crimes. Semen samples
from both victims revealed the blood type of the murderer was the same in
both cases, and other aspects of the crimes strongly suggested a single
assailant. Dr. Alec Jeffreys and his colleagues at Leicester University had
recently developed a technique for creating what are now known as DNA
profiles, and the Leicestershire police contacted him, almost as a last resort.

Jeffreys first compared Buckland's DNA to that of the semen samples.
He proved the police were right, one man had committed both crimes, but
at the same time wrong, since that man was not Richard Buckland.

Independent confirmation of Jeffreys's tests made Richard Buckland the
first person ever to be exonerated by DNA profiling. Good news for Buck-

land, but the police still had two unsolved murders on their hands. They decided to carry out the world's first mass DNA screening. All the adult men in three neighboring villages, about 5,000 in all, were asked to volunteer to provide blood or saliva samples. About 10% of the men matched the blood type of the murderer, and they were all then given full DNA profiles. This took six months—but there were no matches. Then, one year later, a woman came forward to say she had overheard her colleague, Ian Kelly, bragging that he had given his DNA sample under false pretenses. In fact, Kelly had masqueraded as his friend, Colin Pitchfork. Based on this evidence, Pitchfork was arrested and his DNA matched the semen in both victims. Pitchfork was sentenced to life in prison in 1988.

Jeffreys coined the term "genetic fingerprinting," and his technique literally answers the question "Who are we?" Since the human genome inside each of us (and in each of our body's cells) is unique, genetic fingerprinting can be used to identify each of us with absolute certainty (unless you happen to have an identical twin). We speak of "the" human genome, and this is nearly correct—only about 0.1% of our genetic material varies from one person to the next. But that 0.1% still leaves about 3 million base pairs of DNA that differ—more than enough to allow us to identify one another. Only about 5% of the human genome is made up of genes. The other 95% is known as noncoding DNA (or sometimes as "junk" DNA). Although noncoding DNA does not include genes, it does have important functions that are becoming better understood all the time.

The use of noncoding DNA in genetic fingerprinting is a complex combination of molecular biology, biochemistry, and statistics. The technique was first used as evidence in court in 1985. The Pitchfork murder trial was the first time DNA evidence was ever used to both exonerate the prime suspect and convict the real murderer. Since then, the techniques have been greatly improved and streamlined, so that DNA fingerprinting has become more versatile, cheaper, faster, and more reliable.

In some ways, the development and use of DNA fingerprinting mirrors the development of old-fashioned fingerprinting (the kind done on fingers). Among the first uses of traditional fingerprints was simply identifying people—confirming that they were who they said they were (for example, in India, the ruling British used fingerprints to identify Indians who could not sign their own names). The first use of DNA fingerprints was similar. After the basic science had been developed in England, the technique was used to verify the claims of would-be immigrants that they were the close relatives of people who had already immigrated.

In spite of the term "DNA fingerprinting," this genetic technique actually has much more in common scientifically with the technology of blood typing than it does with traditional fingerprinting. The difference, and the reason why the term "DNA fingerprinting" is so apt, is that blood typing doesn't uniquely identify someone, any more than hair color does.

Blood typing is, however, a genetic test—it measures genetic differences from one person to the next. (A single gene on chromosome 9 determines your ABO blood type.) But blood typing is really only an exclusionary test, and not an identification test. If, for example, my ABO blood type does not match that of the blood found at a crime scene, I can be excluded—that's not my blood. But if my blood type does match the blood at the crime scene, this does not mean that it was my blood. There are far too many people with any given ABO blood type to identify someone based on blood type alone. The genius of DNA fingerprinting is that it measures the human genome in ways that are much more restrictive than blood typing. We can use blood type to create an analogy for how this works.

There are four ABO blood groups, A, B, AB, and O. Let's assume (although it isn't true) that these blood groups are equally distributed (25% of all people are type A, 25% type B, and so on). The odds of my having any particular blood type are thus one in four. What if there were a second independent set of blood groups (let's call them CDP), with the four groups C, D, CD, and P, whose presence I could be tested for? Again assume that 25% of all of us have each one of these types. Now, what are the odds that I have any particular combination of ABO and CDP blood types—what are the odds that I am, let's say, both type A and type C? That would be one in sixteen (one in four times one in four). What if there were a third set of blood groups like this? The odds of having any particular combination of all three groups would be one in sixty-four (one in four times one in four times one in four—or one in four to the third power).

A one in sixty-four chance that the blood at the crime scene might belong to someone else is not nearly enough to get me convicted of murder, but what if, instead of three, there were ten different blood tests like the ones described above? Then, the odds that the blood belongs to someone else would be one in four to the tenth power. That's one in 1,048,576—or roughly "one in a million." This is probably still not good enough for a court of law. But if there were fifteen such tests, the odds would be greater than one in a billion. That means that if these fifteen tests were done on a blood sample at a crime scene, and that each of the fifteen results matched my blood, the odds that it was someone else's blood, and not mine, are about one in a billion. Would you convict me based on that kind of evidence?

This is a rough analogy for how DNA fingerprinting works. Multiple tests are performed on snippets of DNA extracted from a sample. (The sample can be blood, hair, skin, bone, or other cells, and, not surprisingly, the details of how the test is done are not at all like a blood test). The results of each test are compared to the results of the same test done on the DNA of a suspect (or, increasingly, they are compared to a growing database of DNA results). The odds that any given test, within the overall DNA fingerprint, will result in a match (like the fictional one-in-four blood tests above) are known. A single DNA fingerprint combines multiple tests like this into one. Thus, when the multiple tests within a single "DNA fingerprint" have been conducted and there is a match, the overall odds that the match could be a different person can be calculated, as in the over-simplified blood test example. With good DNA samples, it is possible to obtain matches with odds in the neighborhood of one in one hundred billion. As a comparison, if I were thinking of an eleven-digit number, say 34,621,986,825, the odds are one in one hundred billion that you could guess that number on the first try. The odds are also about one in one hundred billion that you could correctly guess the results, heads or tails, of thirty-seven consecutive coin tosses.

DNA fingerprinting is a measurement triumph. While it is extremely clever and unquestionably high-tech, it is nonetheless a relatively simple, straightforward application of our accumulated knowledge of the human genome. What's more, it is only the tip of the iceberg in terms of the knowledge humankind continues to develop through the genome.

I, Robot

Many of us are fascinated by robots. The best-known robots are fictional—R2D2 and C3PO of *Star Wars* fame. But real-life robots, such as those welding in a car factory or firefighting ones that go where firemen can't, are admired as well. Once vilified for taking jobs away from people, these robots are now glorified in advertisements and held up as a symbol of intelligent strength and high-tech cool.

While we may admire them, we are at the same time grateful that we aren't robots. A robot is a machine about which everything is measurable; its size, speed, senses (seeing, hearing . . .), computing power, and even its reliability are all transparently knowable, and thus its performance is highly predictable. Human beings aren't like that. Or are we?

As knowledge of the human genome grows, our ability to "measure ourselves" grows ever more profound. Our body type (slender, stout), sexual orientation, tendency to get certain diseases, and so on, may all be encoded to some extent in our genes, and thus measurable. Could this mean that we are all just flesh-and-blood robots?

Leaving aside for a moment the obvious question of "nature versus nurture" (genetics versus environment) in the determination of human characteristics, the human genome allows us to know ever more about ourselves, whether we like it or not.

How long will it be before we have no secrets left, before human beings are no more mysterious than robots? From all indications, that will be one heck of a long time, and it's certainly not clear that we'll ever get there. But things are progressing at a thrilling pace. The Human Genome Project, originally scheduled to be completed in 2005, was completed in 2003, passing most of its major milestones *in advance* of what had once seemed to be an ambitious schedule. How often does that happen on a big government project?

The human genome has been compared to Matt Ridley's book *Genome: The Autobiography of a Species in 23 Chapters*, which is loosely based on the structure of the genome itself. Ridley considers that those of us alive today are truly lucky, in that we shall be the first to be able to "read" the book that is the human genome in its entirety: "Being able to read the genome will tell us more about our origins, our evolution, our nature and our minds than all the efforts of science to date. It will revolutionize anthropology, psychology, medicine, paleontology, and virtually every other science."

We often speak of "the" human genome, but there are as many human genomes as there are human beings. We each have our own. Compared to the overall genome, the differences in the human genome from one person to the next are tiny (but extremely important), and so we speak of one "human genome." But what is it? The human genome is the full complement of "genetic material" in a human cell. Each cell in your body (and there are trillions of them) contains your entire human genome.

It is not my purpose to explain the science of genetics here. I am more interested in how genetics and the human genome relate to our ability to measure ourselves. By that, I mean our ability to understand fundamentally why we are the way we are.

To understand why a robot behaves the way it does, one need only understand the robot's hardware and its software. Those functions, hardware and software, are separate in robots. A robot's hardware is just that: its structure, its means of locomotion, its power supply, and the instruments it uses to sense its environment (that is, to see, hear, feel, etc.). A robot's software is the set of instructions it uses to get the most out of its hardware. A manufacturing robot, for example, might be programmed to move its arm to a precise location, pick up a steel component, move that component to a different location, and then weld it or bolt it into place on the frame of a car. A firefighting robot might be programmed to go into a burning building, search out the hottest location, and deliver a load of fire retardant. These two robots could never switch jobs—both their hardware and software are specialized for their particular tasks. The human genome might be thought of as our "software," but if so, it is a very special kind of software. It does something really amazing: it creates its own hardware.

The genome, be it chicken, human, or whatever, is like a "recipe book." It is a recipe book that can both copy itself and read itself. This gives the genome the two skills possessed by every living thing: the ability to replicate, and the ability to create order. Copying itself is replication; that's simple enough. The ability of the genome to read itself gives it the ability (through a complex process) to create proteins, and proteins either make up or else make nearly everything in our bodies. A fertilized egg is not a chicken, but it can create a chicken. The instruction set, the genome, in an egg has the ability to replicate itself and also to create order, by creating the thousands of different chemicals that go into the hundreds of different kinds of cells the chicken needs.

The Structure of DNA

"Scientists tell me that it is extremely difficult, when one has found something that makes a difference, to recapture the way one thought before," Horace Judson says in his book *The Eighth Day of Creation: Makers of the Revolution in Biology,* and I suspect many nonscientists may have experienced this too. How did we (scientists and laypersons alike) think before we knew about the familiar double helix of DNA and how it accounts for the replication of cells in the body and for the production of all the chemicals that make a person a person, a chimp a chimp, and a fruit fly a fruit fly?

The history of the hunt for the human genome is filled with triumphs of measurement, and with brilliant people brilliantly interpreting those measurements. The most famous moment to date in this fascinating story came in 1953 when James Watson and Francis Crick announced their discovery of the structure of DNA, which is on many short lists of the greatest scientific achievements of all time. On the face of it, what Watson and Crick did (we would say Crick and Watson but for their flip of a coin to determine the order of authorship) does not seem all that spectacular: They simply determined the physical structure (how the atoms of the molecule are arranged in space) of deoxyribose nucleic acid (DNA). In those days, and even today, people were always figuring out the structure of some molecule or other. The structure of DNA, compared to other important biological molecules (such as hemoglobin) is not all that complicated. It's just that the structure of DNA, as Watson and Crick put it in perhaps the most famous understatement in scientific history, "has novel features which are of considerable biological interest." At least, that's how they described it in their landmark April 1953 article in the journal *Nature.* At the Eagle pub in Cambridge, England, just a month earlier, Crick put it a little more forcefully. "We've discovered the secret of life," he said. And so they had.

Watson and Crick's discovery was both an end and a beginning. It was the end of a long search for the means by which genetic material is able to copy itself—the secret of life to which Crick referred. It was the beginning of the search, still ongoing today, for the exact composition and sequencing of all the genes in human beings, and of the search for the understanding of the importance of each of those genes.

The science of genetics was around long before Watson and Crick. It was well known for example that certain diseases were "genetic" or inherited, and that many other traits, such as eye color, were inherited as well. It was even accepted as early as 1944 that DNA itself was intimately involved in the processes of heredity. Before that, most people believed that DNA just played a structural role within the cell—that it was like the poles that hold up the genetic tent, so to speak. At that point, you might say that we were just observing or measuring the results of the action of the human genome, without really understanding what was going on "under the hood."

DNA and the genes it contains have been compared to atomic theory. In many ways, the comparison is a good one. The atom is the basic building block of chemistry. To really understand chemical reactions requires the knowledge that matter is made up of atoms. This was first proposed by John Dalton in 1803.

The "atomic theory" of biology, that is, genetics, owes a great debt to Gregor Mendel's work in the 1860s. Mendel was able to infer the existence of the "atom" of biology, which later came to be known as the gene, from his landmark experiments with peas, flowers, and other plants. When he crossed specific varieties of peas, for example, the resulting hybrids were always like one parent or the other. The "essence" of one of the parents seemed to have disappeared. If he then allowed the hybrid to self-fertilize, the missing characteristic of the grandparent reappeared in one-fourth of the third-generation plants. Thus, the "atom" of that characteristic, be it seed coat color, pod shape, or whatever, really was present in the second generation, even though it couldn't be seen.

We can take the atom-gene analogy one step further. For a long time after Dalton's theory, it was believed that the atom was indivisible. For normal chemical reactions, this is true. But with the discovery of radioactivity in 1895, it became clear that there was an entirely different class of reactions in which certain types of atoms split into other atoms—these are the nuclear reactions, which are the basis for both nuclear energy and nuclear weapons. The genetic equivalent to splitting the atom is mutation. The gene is like an atom. The genes create the chemicals that create the cells that are life. But just as the atom has an internal structure, so does the gene. And that structure can be changed, or mutated. Hermann J. Muller proved this at Columbia University in 1927 by blasting fruit flies with x-rays and then observing that the radiated bugs produced deformed offspring.

The x-rays mutated some of the flies' genes, which then could not produce the right chemicals, and thus certain kinds of cells in the offspring weren't right, and the offspring were defective.

Just because they didn't have Dalton's atomic theory of matter prior to 1803 doesn't mean that scientists weren't able to do some very sophisticated chemistry before that theory was developed. Likewise, a great deal of very sophisticated genetics was developed prior to the work of Crick and Watson.

Measurement was at the heart of Watson and Crick's discovery of the structure of DNA. At the core was the science of x-ray crystallography, invented 40 years earlier, which allows for the measurement of the shapes and dimensions of crystalline substances at the atomic level. X-ray crystallography is complex—even its simplest concepts often give my own students fits. When these measurement techniques are applied to really complicated crystalline structures (for example, a protein such as hemoglobin, which contains thousands of atoms and, unlike the familiar double helix of DNA, does not have a repeating structure) scientists must measure hundreds of parameters from x-ray images in order to determine their structures. By contrast, Watson and Crick worked out the structure of DNA from only three such measurements: the width of the DNA double helix, the height of one complete turn of the helix, and the height from one step to the next up the middle of the double helix. From these, and from cardboard-and-wire models that they built, they were by trial and error able correctly to deduce the structure of DNA where others had failed.

Simply measuring and then understanding the structure of DNA was hardly enough, however, to allow Crick to boast that they had discovered the secret of life. That knowledge had to be linked with a mechanism by which DNA replicates itself, and by which it allows for the formation of the proteins that produce all the different cells in the body. As Max Perutz puts it in Judson's *The Eighth Day of Creation*: "Genes are made of DNA—full stop. The structure of DNA gave to the concept of the gene a physical and chemical meaning by which all of its properties can be interpreted. Most important, DNA—right there in the physical facts of its structure—is both autocatalytic and heterocatalytic. That is, genes have the dual function, to dictate the construction of more DNA, identical to themselves, and to dictate the construction of proteins very different from themselves."

Watson and Crick discovered the structure of DNA, and, knowing that DNA was at the heart of heredity, suggested how the structure of DNA would allow it both to reproduce itself and also to produce things like hemoglobin and insulin—the chemicals of life.

The Genome Today

The Human Genome Project, which began in 1990 and officially ended in 2003, was a broad, complex undertaking, and perusal of its ambitious goals reveals a measurement theme. For example, an early goal was to "complete a genetic linkage map at a resolution of two to five centimorgans by 1995." (As with the other project goals, this one was met ahead of time: by 1994, the genome had been mapped to a resolution of 1 centimorgan.) One centimorgan, or cM, is equal to about one million base pairs of DNA in human beings, which, very roughly, might be about a half a millimeter. The base pair is the smallest unit of measure along a strand of DNA. Human DNA is several billion base pairs long.

By April 2003, more than 99% of the gene-containing portions of the human genome had been sequenced, with an accuracy of over 99.99%. In addition, the genomes of several other species had been sequenced, including several bacteria and a fruit fly. Draft versions of the mouse and rat genomes were also produced.

In many ways, though, the work is just beginning. We are entering, some have said, the "genomic era." Change will be fast and furious. The implications for biology, human health, and society are more profound than almost any of us can imagine.

Who Were We, Who Are We, and Who Are We Going to Become?

The human genome is itself a work in progress. Genes change, or mutate, from one generation to the next. As we learn more and more about the human genome as it exists today, we are able to infer amazing things about how it got that way. DNA is the recipe for life, and we are understanding more and more about how that recipe has changed over time. That understanding helps us to predict how and when future mutations will change the genome, and thus life itself.

It is hard to resist the analogy of the computer program at this point, overly simplistic (and in some ways misleading) though it may be. The genome is like an impossibly long, incredibly complex computer code. Computer programs evolve. New sections of code are added to perform new functions, increase user-friendliness and reliability, resist viruses and hackers, and so forth. At the same time, redundant, inefficient, or obsolete sections of code are weeded out. We understand (well, sometimes anyway) why changes are made to computer programs. We don't always understand everything about the processes of genetic change yet.

Nonetheless, genomic research gives us completely new ways of investigating the past (the origins of our species), understanding the present (why we look and behave the way we do), and predicting the future (what our children will be like, their children's children, and so on for thousands of future generations). We can investigate various aspects of the past, present, and future in other ways, but our growing knowledge of the human genome allows us, in many cases, to get to the very core of important questions, instead of just poking around the edges.

Who Were We?

If we, as a species, want to understand our past, there are many different techniques—many different formalized ways of learning—available to us. First, there is the historical record, consisting of books, drawings, paintings, and, more recently, photographs and audio and video recordings. Then there are archaeological remains. Digging up what's left of ancient (and not so ancient) civilizations teaches us how they lived and what was important to them. Linguistics, too, can offer clues as to how prehistoric humans developed, migrated, and populated the earth, as we learn more about the thousands of human languages, most of them now "dead," and their common roots. Going further back in time, the fossil record can help us understand the mysterious origins of the human on earth, and of our relationship to species that still exist, such as gorillas and chimpanzees, and to others that are now extinct. These are just a few examples. Historians, archaeologists, linguistics experts, and anthropologists, among others, have all contributed much and still have a great deal more to offer our insatiable curiosity about our past.

Add to those the entirely different set of tools for examining the past that come from knowledge of the human genome. Knowledge of the genome is sweeping in its ability to teach us about our past. Genetic material has been present in every living thing that has ever existed on this planet. It is the one common "language" of life. All of our other tools for exploring the past are limited. Historians are limited by what was written, drawn, painted, photographed, and recorded (and within that, by what has survived). Archaeologists and anthropologists are similarly limited by what they can dig up. Even the fossil record does not stretch back, it would seem, to the very beginning. Those rocks are gone—they have melted and been reshaped. The genome is the one common source of information about life that encompasses all of the history of life, right back to the birth of our planet.

Perusing some of the mountains of research devoted to this subject, I can't help but think about how a curious child might react upon receiving the gift of a microscope. The child will immediately begin examining everything she can get her hands on—bugs, dirt, food, fingernails, and so on—with this marvelous new tool. Everything looks different under a microscope. And so it is as we expand the uses of our new genetic tools to look at our species and how it got where it is today.

More and more detailed knowledge of the human genome will give us knowledge about our species in ways we can scarcely imagine. By comparing our genome to those of animals to whom we are genetically similar, we'll be able to quantify when our species first separated from other primate species. The term "similar" here bears defining. Humans share roughly 98% of their genome with chimpanzees. The genome is a "book" with one billion words—as long as 800 Bibles. That we are 98% like chimps means that out of every 100 "letters" in that book, on average 98 of them are the same for humans and chimps. That also means, though, that there is 16 Bibles' worth of difference.

Most (but by no means all) people have become reconciled with the fact that chimps and humans have common ancestors. Addressing the Pontifical Academy of Sciences in 1996, Pope John Paul II even argued that there was a point in the development of our species, somewhere between the ancient apes and today's humans, at which God inserted a human soul into an animal—a so-called "ontological discontinuity."

But what about the common house fly? How does it make you feel to have ancestors in common with that lowly (not to say disgusting) creature? Scientists routinely perform "gene rescue" experiments that show just how much like flies we really are. In a gene rescue experiment, a gene in one species (the fly in this example) is destroyed by mutation and then replaced with the equivalent gene from another species (in this case a human being). In this way, normal flies can be grown that contain human genes, such that it is frequently impossible to tell whether a fly was "rescued" with a human gene or a fly gene. Geneticists tell us that the last common ancestor between flies and humans existed over 500 million years ago. Yet pieces of our genetic code will still run on a fly's "operating system." If only modern software were so interchangeable.

How do we know, by the way, that humans and flies share a common ancestor that existed over 500 million years ago? The rate at which genes accumulate changes helps us to understand the relationships between the species. For example, from what is known so far about differences in their respective genomes, it is likely that the ancestral line of the human species split from that of the chimpanzees sometime between 5 and 10 million years ago. Genetic time, however, is better measured in generations, not years. Ridley invites us to envisage a line of 300,000 women linking hands from New York City to Washington, D.C. Each woman stands behind her daughter and in front of her mother, with the youngest of the women in New York. If there were a similar line of 300,000 or so female chimpanzees parallel to the line of women, the two lines would join in Washington. That is the approximate point in genetic time when the two species split.

Who Are We?

From a genetics perspective, we can look at this question in at least two ways. The first concerns anatomy and behavior. Our genes have a lot to say about what we look like (our hair color, size and shape, and so on). Behavior is more controversial, but there is a consensus that our genes have a huge influence on the type of person we become. The second way in which genetics answers the question "Who are we?" is the literal answer—the one provided by DNA fingerprinting.

DNA fingerprinting is not especially controversial. Even lawyers, it seems, have come to terms with it. However, things are not quite so simple when we consider the other ways in which the human genome helps us to understand who we are.

That genes have a great deal to say about our anatomy is generally agreed upon. There may not be one single "obesity gene," for example, but so much of what we look like is hard-wired into our genes that it would be difficult to provide a comprehensive list. That genes also exert some control over our behavior, in ways we are only beginning to understand, is just as true, although much more controversial.

The genome changes—but it can only do so from one generation to the next. For our species (or any of the others) to be able to make shorter-term changes, within a single generation, another mechanism (besides gene changes) is needed. That's where our brains come in. This is the balance between instinct and learning. Much about us is instinctive. We instinctively pull back when we touch something hot, we instinctively nurse at our mother's breast, and much of learning to speak is instinctive. On the other hand, we must actively learn how to read, how to dress ourselves, and how to drive a car. Those things are not hard-wired into the genetic code. One could even imagine that the genome, finding itself unable to make short-term improvements within a single generation, evolved the brain as a means of handling these things.

The question of instinct versus learning becomes more complicated when we realize that some instincts require outside information—that they must be, in some sense, learned. There is much evidence that grammar is instinctive—there are unifying grammatical themes across all languages that suggest a universal grammar. Children consistently apply complex grammatical rules without having been formally taught—they just seem to know. Yet children who are not exposed to language at an early age do not "express their grammatical instincts." When they are older, these children can learn grammar, but only through laborious study.

Anyone who has studied a foreign language after childhood knows there is something to this business of grammatical instincts. A friend and colleague of mine arrived in the United States from his home in eastern Europe in his late 30s, having never spoken English. As a physicist, however, he read and wrote English relatively well (English being the professional language of physics), and he had written and published scholarly papers in English-language journals. But he was forced to learn to speak English for the first time when he arrived here, and it was a struggle. Eventually, he could communicate, but his atrocious grammar and Ivan the Terrible accent were almost as comical as they were difficult to understand. A year after his arrival, his two children, aged 5 and 7, arrived with their mother. The kids stepped off the plane bereft of any knowledge of English—they did not know a single word of it. During the first few weeks, the kids were very distressed. But they picked up English with astonishing speed and facility. Within a few months, they spoke fluent English with an American accent and had begun correcting their father's grammar—much to his chagrin. They were even embarrassed to be with him in public if he had to speak. "Let me buy that, Dad," his son would say, when they stopped at the market for a loaf of bread.

There have been and doubtless will continue to be many studies that attempt to link behavior and genetics. There are fascinating connections between stress and genetics, for example. Subtle, imperfectly understood genetic differences separate those of us who tend to be perpetually stressed-out from the mellow, laid-back types. Likewise, there are said to be genetic links to, among other things, risk-taking, sexual promiscuity, homosexuality, and shyness.

Those who study genetic links to behavior have before them the intelligence story. It is ironic, as discussed in the previous chapter, that attempts to study the heritability of "intelligence" have been characterized by so much "stupidity." But given the proper perspective, it is not hard to see that genetics research can help illuminate human behavior. Genetics helps complete the picture. We have genes, which control the production of various chemicals. Some of these chemicals (testosterone, cortisol, and so on) clearly influence behavior. At the same time, we have a brain—a highly optimized learning machine. My particular genes may produce chemicals that make me want to take risks, but my brain has learned enough to tell me, "Hey, bungee jumping is stupid—let's go play tennis instead." The brain does not completely control the genes, any more than the genes completely control the brain, however. Throw into the mix healthy doses of cultural and parental influences, add food (or its lack), and stir briskly. You get some idea of how complicated we are.

Some very smart people, not that long ago, thought that intelligence was determined by how big your head was. We've come a long way. Yes, there are genes that have some influence on intelligence (and many other types of behavior), but so do education, diet, and a variety of other influences going right back to the very substantial contributions of the time we spent in the womb.

The competing influences of genetics and environment on a whole host of factors are only beginning to be understood. In terms of "who we are," it's not just behavioral factors, either. Many of us would probably guess that the influences on something like a person's height are most likely to be overwhelmingly genetic and not environmental. But the people who study historical trends in human height (there are such people, and they are very serious about it) will tell you that there are enormous environmental influences on height. In Chapter 5, the relentless improvement in quantifiable sporting records was compared to similar improvements in IQ measurements. If you guessed from those results that the same thing would be true with respect to historical records of human height, that is, that we would be getting ever taller, you would be wrong. Burkhard Bilger describes the research of "anthropometric historians" such as John Komlos. It turns out that historical records of the heights of soldiers in various armies (mostly in Europe) go back many centuries. These records have been compared to those of modern soldiers in those countries. The results show that, "in Northern Europe over the past twelve hundred years human stature has followed a U-shaped curve: from a high around 800 A.D., to a low sometime in the seventeenth century, and back up again."

Among Komlos's other findings is that, while northern Europeans in general continue to get taller (the Dutch are the world's tallest people), the average height of Americans has remained flat for about 50 years. The average American man (not including immigrants, Asian-Americans, or Hispanic-Americans) is five feet nine and a half inches, while the average Dutchman is six foot one. On average, American men are less than one inch taller than the average American soldier during the Revolutionary War. The height of the average American woman has even decreased over the past 50 years.

All kinds of explanations have been suggested for these counterintuitive results, most of them centering on economic or dietary factors, and I do not wish to add my own two cents' worth to these. I have, however, two reasons for describing these results here. First, there is the question of the influence of genetics versus environment. As noted above, many of us have the vague notion that the human race is relentlessly getting better: smarter, faster, stronger, longer-lived, and so on. We further believe that genetic and environmental factors combine to ensure this steady improvement. Generation by generation, better genes (smarter genes, faster genes) will win out over inferior ones, or so one would think. That, combined with perceived environmental improvements (health care, diet, education, and so on) ensures the steady improvement of the human race—thus we should be getting ever taller. If they do nothing else, the historical height data give us pause to rethink these conclusions.

The second reason for discussing the historical height data here is the "to measure is to know" argument. Many of us have been to museums or old castles where we've seen and laughed at the tiny suits of armor men wore back when knights were bold (and short). Anthropometric historical data replace those meager and unsatisfactory notions with the quantitative knowledge of how our average heights have changed over time—leaving us better informed and in a better position to explain why.

Who Are We Going to Become?

We can take the long view or the short view on this question. The short view concerns our children. Who will they be? Genomic research has a lot to say about that, and much of it is quite controversial. The long view concerns where our species is headed, as determined by the changes in the human genome that are in store for us one hundred, one thousand, or one million generations from now.

Let's take the short view first. A child inherits half of its genetic material from its mother and half from its father. The more we know about the genetic makeup of the parents, the more we know or can predict about what the child will be like.

This knowledge has been applied in a variety of situations. The Committee for the Prevention of Jewish Genetic Diseases, for example, runs a program for testing the DNA of Jewish children. The results are compiled in a database that can be consulted later by Jewish couples considering marriage. Both members of the couple can thus learn if they carry the same mutation of a gene that allows a certain disease known to flourish in the Ashkenazi Jewish community (diseases such as cystic fibrosis, Tay-Sachs disease, or a number of others). Thus, if both members of the couple do carry the same such gene, they have the option to either decline to marry or to refrain from having their own children if they do marry. Not everyone is wild about this. In 1993, the *New York Times* called this program "eugenic." On the other hand, cystic fibrosis has almost disappeared from the American Ashkenazi Jewish population as a result of this testing program.

What do you think about this example? One way or the other, you had better think about it, because situations like this are going to become more and more common. By "situations like this," I mean situations in which genetic knowledge (or the potential to have such knowledge) about ourselves or our offspring will present us with difficult choices. Dealing with issues such as this is, not surprisingly, a formal part of government-sponsored genomic research. These areas of research have even been given their own acronym—a sure sign of their importance, at least in the eyes of our government. ELSI stands, not for bovine genetic research, but for ethical, legal, and social issues related to genomics research.

Among the most common uses of the genetic screening of fetuses is for Down syndrome. Older mothers in particular are at risk of producing a child with Down syndrome (the odds are 1 in 100 for a 40-year-old mother but only 1 in 2,300 for a 20-year-old one). Children with Down syndrome are born with an extra copy of chromosome 21. They are generally healthy and happy people. But they are destined to die before the age of 40, usually of a disease like Alzheimer's, and they are mentally retarded. A pregnant woman can know with absolute certainty if she will give birth to a baby with Down syndrome. Would you want to know? Would you abort a fetus with Down syndrome or allow it to be born? Get used to thinking about questions like this, as all pregnant women must do these days. Some related applications of genetic knowledge are far more easily condemned. Among the most popular uses of amniocentesis in India is for determining the gender of a fetus—frequently, so that it can be aborted if it is female.

The long view of who we shall become is more complex, but it is becoming clearer all the time. The human genome has been sequenced, opening the door for a great many extremely important, monumentally complex challenges. The functions of most of the genes are certainly not well understood, and the genes themselves, the regions within the genome that contain the recipes for building proteins, are only 1 to 2% of the overall genome. The functions of the rest of the genome are even more poorly understood.

Even so, we have learned a great deal about who we were, who we are, and who our children will be. What about predicting the future, beyond our children? Lots of things are always going on inside the human genome. The genome is "alive" and ever-changing in genetic time, from one generation to the next. Would we be able to mathematically model and thus predict changes in the genome if we had a map of it and an understanding of each gene's function? How many generations will it be till we as a species lose our fifth toe? We can predict the weather, with some success, using sophisticated mathematical models coupled with measurements from satellites and other sources of data. The genome, in some ways, is less complicated than the weather. Shall we someday be able to make accurate predictions in "genetic time," generation by generation?

Competition among our genes is fascinating. It could even become a popular spectator sport, but for the fact that it only takes places in genetic time, from one generation to the next. One subplot in the story of genetic competition concerns a patch of DNA near and dear to my heart, the Y chromosome. The presence of the Y chromosome is what makes a mammal male. It is the smallest of the twenty-three chromosomes, and in genetic terms, it is in trouble.

In mammals, the female is the "default" sex (in birds it is the male). Thus, in mammals, a male might be thought of as just a modified female: a single gene on the Y chromosome turns on the changes that transform the genitals and reshape the body, leaving behind traces of femaleness such as the vestigial male breasts. Aside from this powerful gene, most of the rest of the Y chromosome consists of noncoding DNA, and the overall length of the Y chromosome is shrinking as it sheds any nonessential genetic sequences. The Y chromosome appears to have circled its wagons and is under attack from all sides.

As a red-blooded male, I find this news disturbing. Having had to put up with us for all these generations, will women finally have the last laugh? Will genetic modeling be able to accurately predict the outcome of this (and many other) conflicts in the battleground that is the human genome?

The Last Word

The collective results of research on the human genome are frequently viewed as a two-edged sword. The benefits are enormous. The sequencing of the genome will almost inevitably lead to better treatments for everything from horrible diseases like cancer to conditions such as chronic overeating. It should also greatly reduce the time and expense necessary to get a new drug to the marketplace, which requires, on average, 14 years and 500 million dollars. The ability to look at the genes of the individual subjects who will test out experimental drugs should reduce the problem of adverse reactions that slows the progress of clinical trials of such drugs.

The ethical, legal, and social implications of research into the human genome are the other edge of the sword. Understanding the underlying genetic causes of thousands of diseases (sickle cell anemia, Huntington's disease, cystic fibrosis, and many forms of cancer, for example) will make us much better at predicting the likelihood that these diseases will occur in any given individual. However, the possibility that these new abilities could be abused, by employers, for one example, is clear.

There are also a great many unsolved issues regarding how the various results of the Human Genome Project might be patented and commercialized. This means more work for lawyers, presumably, but also for educators charged with effectively teaching us about genetic research and what it all means.

Imagine what it would be like if a great genius from the past, a young Isaac Newton say, were shown a modern desktop computer. He would surely be fascinated beyond words by all kinds of things—the fancy audio and video effects and all that, and even something as mundane as word processing—but think of it: here is a small machine that can perform symbolic mathematical operations, right on the screen. Newton invented the calculus. He would be watching a machine that could bring his invention to life! Surely he would be wild to know how such a machine worked. Without any sort of instruction (and not knowing how to google), he would be left with making observations "at the surface" in order to learn more about the essence of the machine. And so it has been for us throughout most of history as we have attempted to learn more about our species. Now, at last, we have simultaneously gotten inside both the motherboard and the operating system and are beginning to dissect both the central processing chip, atom by atom, and the operating system code, line by line.

CHAPTER 8

IT'S
NOT
JUST
THE
HEAT,
IT'S
THE
HUMIDITY

Global Warming and

Environmental Measurement

There is perhaps no beguilement more insidious and dangerous than an elaborate and elegant mathematical process built upon unfortified premises.
—T. C. Chamberlain (1899)

Spaceship Earth

We can never know the true value of anything that we measure. If I weigh myself on 10 different bathroom scales, I'll get 10 different results, none of which is my "true" weight. My true weight is forever unknowable (which is just as well, perhaps). There are, however, a variety of ways of estimating the error in any of those bathroom-scale measurements—or in any other measurement. "Error" in measurement is the difference between an individual measurement and the true value of the thing being measured (the true value being the one we can never know). Thus we can never know the true amount of error, either, but we can do a good job of estimating it. All of this is the sort of thing that one learns about in a formal course on measurement, which most scientists and engineers study in one form or another as undergraduates.

Error is important in anything we measure, but it is extremely important when it comes to the environment. This is because the measurements (such as those associated with global warming) are complex and difficult to perform, and because the conclusions we draw from those measurements, and the actions that those conclusions might precipitate, are so important to us.

Environmental issues are in the news every day, probably none more so than global warming. As I write these words in 2005, global warming remains the environmental story *du jour.* All it takes is a summertime heat wave for everyone to ask: Is our planet getting warmer? This question leads directly to others, even more profound: If our planet is getting warmer, to what extent is this warming trend man-made? What are the likely results of global warming? If, indeed, humans are to some extent responsible for any global warming trends that exist, what can be done about this? To begin to answer any of these questions properly pushes measurement technology to its very limits, to say nothing of the demands it places on our collective wisdom in interpreting and then acting on the results.

Vital Statistics—the Age of the Earth

We all know approximately how much we weigh, how tall we are, and how old we are. Even very young children can answer these questions about themselves. Why shouldn't we be curious about these same vital statistics regarding our home the Earth?

Measuring the age of the Earth is complicated. The Earth has no birth certificate, and as far as we know, there were no witnesses to the blessed event (at least none that are still around). Measuring how long Mother Earth has been around is a thorny problem. Let's first consider a much simpler problem: How would you go about measuring or otherwise determining your own age, if you didn't already know it? Let's say you had a terrible case of amnesia, like the character portrayed by Guy Pearce in Christopher Nolan's movie *Memento* (2000), and you couldn't remember anything about yourself. Figuring out your age might be the least of your problems, but what if you really needed to know it? What could you do?

There are many techniques you might employ. Some are rough approximations. You could employ a process of comparison: if you looked in the mirror and saw gray hair, a receding hairline, and a wrinkled face, you might conclude you are roughly the same age as acquaintances that have similar features, and whose age you knew or could find out. If you needed a better approximation, a dentist could analyze your teeth and give you a reasonable estimate of how old you are. Likewise, an ophthalmologist might examine your eyes and do the same thing. Other medical or scientific professionals could help in similar ways.

You might also be able to infer something about your age from other circumstances. For instance, if you found out that you had a 25-year-old child, you would be pretty safe in concluding that you are at least 40, and probably older. You might also infer your age from things that you found in your house (assuming it really was your house). The date on a college diploma, for example, would be useful. Old letters or photographs might help, too. Even your collection of recorded music or your wardrobe could yield significant clues.

For any one of these scientific estimates or circumstantial inferences of your age, you would of course want independent confirmation. It would be nice to hear a doctor tell you that you don't look a day over 30, but that would be contradicted by the existence of your 25-year-old child! In the end, however, you could probably arrive at a pretty good estimate of your age and confirm it independently in several ways. Most people have concluded that this is just what we have done with the age of the Earth, whose age is overwhelmingly (but not universally) agreed to be about 4.5 billion years, plus or minus 1%. We should all look so good at that age.

There is a rich literature and long history related to the measurement of the age of the Earth. One might wonder why we as human beings are so interested in this question. The Earth's age does not influence public policy. That was not always true, however. The question of the age of the Earth was for some time linked—by Lord Kelvin and others—to the question of how long the Earth might be expected to go on living—a serious "public policy" question indeed!

We care about the Earth's age because of plain old curiosity. In addition, the age of the Earth has serious religious implications. Depending on how one reads it, the Bible can be construed to imply that the Earth is relatively young—perhaps only thousands of years old. In the early 1600s, Archbishop James Ussher of the Anglican Church produced a meticulous study that worked its way backwards through the biblical genealogies to arrive at a creation date of 4004 BC. Today, many creationists are interested in scientific measurements to ratify their religious belief that the Earth is indeed quite young. Moreover, if the Earth could be scientifically shown to be young, Darwinian evolution would be discredited, because one of its linchpins is that all the plant and animal species on the planet today have evolved to their current state over a large number of generations. This would not be possible if the Earth's age could be measured in just thousands, or even millions, of years. So the age of the Earth, rather than being just a simple matter of curiosity, has serious scientific, religious, and cultural implications.

Measuring the Age of the Earth

Humankind has been interested in the age of the Earth for at least 2,000 years. Prior to that, some ancient civilizations believed that the Earth had existed "forever." Western thought on the question of the age of the Earth remained mostly religious until about 1860. Between 1860 and 1930, a vigorous debate involving studies in physics, geology, astronomy, and biology took place, with a variety of different approaches being attempted to determine the Earth's age through measurement or calculation. Whatever branch of science was used, some phenomenon had to be identified that could be accurately measured, and that could be shown to vary regularly with time. Today, the decay of radioactive elements is regarded as the only phenomenon that meets these criteria. Prior to the discovery of radioactivity in 1903, other criteria had to be employed. During the 1800s, these various approaches fell into three major categories: physical, astronomical, and geological chronometry.

Kelvin and the Physical Chromometers

The first of these three categories is physical chronometry, which applied thermodynamics and other laws of Newtonian physics to the problem. Lord Kelvin was the best known of those using these techniques. He calculated the amount of energy available to both the Earth and the Sun from all sources (save the most important source of energy, nuclear energy, which had not yet been discovered). Then he measured or estimated the rates at which each body dissipates energy to determine the time needed for each body to cool from its assumed original state to its present condition. A coroner might use a similar approach to estimate the time of death of a murder victim—provided the victim had been dead for only a short time—not long enough for the body to have cooled off uniformly to room temperature.

It was well known even before Kelvin's time that the Earth's temperature increases with depth—the closer you are to the center of the Earth, the warmer it will be. This suggests that the Earth is losing heat through conduction from its hot core to its cooler surface. Kelvin concluded that the Earth had once been a sphere of molten rock, and that it had solidified from the surface inwards. After the Earth finished solidifying, it would continue cooling until it became a solid sphere of uniform temperature, he further concluded. His approach employs thermodynamic principles to calculate how long it took the Earth to reach its temperature at the time of his study. To do so, he needed to measure or estimate the Earth's internal temperature, the temperature gradient at its surface, and the thermal conductivities of the Earth's constituent rocks. Data were (and are) available to show that, near the surface, the Earth gets warmer by about 1°F for every 50 feet of depth. Kelvin also convinced himself that he had accurate enough measurements of the thermal conductivities of the rocks he believed were representative of those found throughout the Earth. As for the temperature of the Earth's core, he assumed this to be the melting temperature of the constituent rocks. Kelvin was aware from the beginning that there were significant uncertainties in the measurements he used, to say nothing of those inherent in his assumptions of the initial conditions of the problem. When all was said and done, his final calculation resulted in an age of the Earth of 98 million years. Because of uncertainties in the various data, he determined upper and lower bounds of 20 million years and 400 million years.

Kelvin's major work on the subject, "On the Secular Cooling of the Earth," was not produced simply to provide an answer to an interesting scientific question. (Kelvin employs a less familiar meaning of "secular" here: "taking place over an extremely long period of time.") Kelvin was candid in his desire to show the errors in the ways of geologists, among others, who believed in a geological theory known as uniformitarianism. Briefly, this theory states that current geologic processes, occurring at the same rates observed today, in the same manner, account for all of Earth's geological features. The idea that Earth was thus in a sort of "steady state" appeared to Kelvin to violate the laws of thermodynamics. If the Earth were continually, if slowly, cooling down, then a theory that postulated steady state processes must be invalid. With his measurements and calculations of the age of the Earth, Kelvin wished to show that the geologists of his time were ignorant of the most basic of scientific principles.

Kelvin's real contribution to the age-of-the-Earth question was not so much his approach, innovative though it was, or his results, which were seriously flawed. What he did contribute was to show that, to the extent that we understand the laws of nature, they can and should be applied to scientific questions of any type or magnitude. Prior to Kelvin and some of his contemporaries, even scientific investigations into the age of the Earth were heavily influenced by religious beliefs. Kelvin, a man of deep religious faith, simply said that the age of the Earth is a scientific question and that it can be answered using the same laws of science that we would use to answer any other scientific question.

There were many problems with Kelvin's approach to the measurement of the age of the Earth, although in fairness to Kelvin, one of the most important problems was one he could scarcely have anticipated. That is the presence of vast quantities of radioactive elements in the Earth, which are constantly giving off tremendous amounts of heat as they decay. The Earth thus appeared to Kelvin warmer than it should have been, and he estimated its age as much younger than it actually was. (The same problem obstructed Kelvin's attempts to determine the age of the Sun.)

Astronomical and Geological Chronometry

The second major set of approaches to measuring the Earth's age in the nineteenth century, which involved measuring anomalies in the Earth's motion through space, could be called astronomical chronometry. Two such anomalies are most often cited. One of these is that the Moon is ever so slowly accelerating away from the Earth. If, as most suspected, the Moon had once been a part of the Earth and had been flung off when a young Earth was still molten, it might be possible to calculate from measured changes in the Moon's motion how long ago it must have been separated from the Earth. This would not give the Earth's total age, but it would give a nice lower boundary to that age.

The other frequently cited anomaly is that the ellipticity of the Earth's orbit around the Sun changes with time. If the Earth orbited around the Sun in a perfect circle, the seasons would be of equal length, at least at the equator. The more elliptical the Earth's orbit is, the greater the difference in the length of the seasons will be (an ellipse is a sort of stretched-out circle). It was postulated that periodic changes in the Earth's climate (such as ice ages) might be related to these changes in orbital shape. Since astronomical calculations were able to date epochs of high orbital eccentricity, it was believed by some that the dates of previous ice ages on Earth could be established, and that these could be further extrapolated to the dates of even earlier ice ages. Once again, this does not give the actual age of the Earth (even if all the measurements, calculations, and assumptions are accurate), but it does shed some light on it. (This approach is also related to the question of global warming—since explaining when and why the various ice ages came and went is an important part of understanding the history, and predicting the future, of our climate.)

A third set of approaches to measuring the age of the Earth in the nineteenth century involved geological chronometry. In the first half of the nineteenth century, geologists had drawn up a careful record of the Earth's history as a function of the forms of life that had existed over time, as revealed in the successive layers of the Earth's crust. This was extremely useful for several purposes, but it wasn't much help in determining the Earth's age, since it wasn't known how long it had taken for each layer of the Earth to be deposited. To use this approach, it was necessary to determine, by studying rates of erosion and sedimentation, how long it had taken for the various layers in the Earth's crust to form. As with the physical and astronomical approaches, this approach required broad assumptions, tedious calculations, and careful measurements. There were a great many advocates of this approach, however, and they generated an enormous amount of geological data. Surprisingly, their estimates for the Earth's age, while numerous, did not vary all that much from Kelvin's best thermodynamic estimate of 98 million years.

Charles Darwin himself attempted to use sedimentation measurements to estimate the age of the Earth. Darwin's theory of evolution of necessity required that the Earth be quite old, because of the tremendous amount of time necessary for the natural selection of the species to take place. Darwin's attempts to quantify the age of the Earth, however, were disastrous. Early editions of his landmark book *On the Origin of Species* contain a hastily done calculation of 300 million years, based on sedimentation measurements. His critics, Kelvin prominently among them, pounced on weaknesses in his approach, and Darwin's misstep threatened to detract from the brilliance of much of the rest of his masterpiece. (The section containing the calculations of the age of the Earth was removed from later editions of Darwin's book.) Despite his inability to quantify the age of the Earth, Darwin made some qualitative contributions to the discussion regarding the age of the Earth through his theory of natural selection and the evidence he presents for it, in the fossil record and elsewhere.

Other schemes for measuring the age of the Earth were developed as well. Much effort was expended, for example, in developing a "salt clock," which used the level of salinity in the ocean combined with the rate at which salt was being added to the ocean by the rivers of the world. The most significant problem (among several) with this is that we now know that all the sodium added to the ocean does not remain there (as was assumed by those applying this technique), but that much of it evaporates and is recycled via rain. The "salt clock" can thus be thought of as a means of estimating how long sodium remains in the ocean on average, and not the age of the Earth.

Radiometric Methods

The discovery of radioactivity late in the nineteenth century eventually changed the entire debate about the age of the Earth. One immediate effect of this discovery on the age-of-the-Earth problem was the realization that nuclear reactions were an obvious source of energy not just inside the Earth but for the Sun as well—and that thus Kelvin's calculations of the Sun's age, and also when it would run out of fuel and energy, were greatly in error. Based on the nuclear energy argument, the Sun was much older and had enough energy to last much longer than Kelvin had predicted.

Radioactivity became a far more accurate "geo-chronometer" than any of the other techniques described above. Following the initial discoveries of Henri Becquerel and Wilhelm Roentgen (he for whom the basic measurement unit of radiation is named) in the 1890s, there followed a rapid series of advances in the physics and chemistry of radioactivity. Marie Curie and her husband Pierre, Ernest Rutherford, and others did the basic experiments and formulated the theories on which several entire industries (nuclear energy, nuclear weapons, and nuclear medicine leap to mind) depend today. It was Rutherford, along with Frederick Soddy, who in 1902 first formulated a general theory of the rates at which radioactive elements decay into other elements. Rutherford and Soddy isolated a radioactive gas and measured the gas's radioactivity versus time. After 54.5 seconds, the radioactivity of the gas was half of its original value. In twice that time (109 seconds), the radiation level had been cut in half again. In fact, for every 54.5 seconds thereafter, the level of radiation again diminished by half. Similar experiments on other radioactive gases showed the same thing: the radiation diminished by half, "like clockwork," after a constant period of time—the only difference being the time required. Different radioactive gases had different time constants. These and related experiments led to the realization that radioactive elements are unstable and that they decay spontaneously to form other elements, all the while emitting radiation.

In 1905, Rutherford suggested that the traces of helium gas found in radioactive minerals were probably there because of the decay of radium and other radioactive elements in those minerals. Measuring the amounts of helium and of the radioactive elements in the rocks, and knowing the rate at which helium was produced by those radioactive elements, it should therefore be possible to determine the age of the rocks. Helium, though, is a gas, and over time, some of it is likely to escape from the rock whose age is in question. Rutherford later went on to suggest that age measurements based on the production of lead, a solid, by radioactive decay, instead of gaseous helium, might prove superior. He was right. The technique that eventually yielded the most widely accepted values for the age of both the Earth and the solar system involved the decay of uranium into lead.

By 1927, the rate of decay of uranium into lead was well enough understood for Arthur Holmes of the University of Edinburgh to include it in a table showing various estimates of the age of the Earth based on different methods of calculation, reproduced here as Table 8.1. Most of the techniques discussed earlier in this chapter appear in Holmes's table, with the notable exception of the work of Kelvin and others on cooling models, which had been so thoroughly discredited by 1927 that Holmes felt justified in omitting them. Note also that although Holmes included geological sedimentation thickness in his table, he concluded that the age of the Earth was "incalculable" from such measurements.

Holmes reported the age of the Earth as less than three billion years, based on the uranium and lead "in average rocks." As noted earlier, this figure has since been revised upward; the most widely accepted value today is 4.54 billion years. The change is the result of advances in instrumentation, in theory, and in technique. Looking around a lot hasn't hurt either. Geologists have left few stones unturned in their search for older and older rocks. Rocks exceeding 3.5 billion years have been found on all the Earth's continents. Meteorites are the oldest rocks available to geologists on Earth. Extensive testing by a suite of radiometric dating techniques shows that many of these rocks are around 4.5 billion years old. Moon rocks are nearly always extremely old. Since the Moon's geological history has been quiet compared to that of the Earth, Moon rocks between 3.5 and 4.0 billion years old are common. If one believes the currently favored hypothesis that the Moon was formed after a planetoid crashed into the Earth, then the age of Moon rocks provides a check on the age of the Earth.

No measurement is perfect, and no scientific theory is beyond revision.

Table 8.1

Various estimates of the age of the Earth as of 1927

Method	Estimated age of Earth (billions of years)
Eccentricity of the orbit of Mercury	1–5
Tidal theory of the origin of the moon	< 5
Journey of solar system from the Milky Way	2–3
Uranium and lead in average rocks	< 3
Oldest analyzed radioactive minerals	> 1.4
Sodium in the oceans	$n * 0.3$ (where n is a constant, such as 5)
Thickness of geological formations	incalculable
Cycles and revolutions in Earth's history	> 1.4

Source: Adapted from G. Brent Dalrymple, *The Age of the Earth*, © 1991 by the Board of Trustees of the Leland Stanford Jr. University, by permission of the publisher.

As G. Brent Dalrymple notes in his 1991 book *The Age of the Earth*, "scientific conclusions are always tentative, and the age of the Earth is subject to revision should new evidence require it." But at the same time, he concludes: "We can be confident that the minimum age for the Earth exceeds 4 billion years—the evidence is abundant and compelling."

Global Warming—Taking the Temperature of the Earth

Climate change, in particular global warming, has become the most reported-on and fretted-about environmental concern since the early 1990s. Ironically, most of us seem to have forgotten all the sober predictions of a new ice age that were in vogue only twenty to thirty years ago. However, whether the future holds global warming, a new ice age, or something else entirely is extremely difficult to predict. A better question to ask is whether human influences are disturbing the complex interactions among our atmosphere, our oceans, and our land masses in ways that are likely to have negative consequences.

The measurement of global warming has at least one thing in common with the measurement of intelligence. Both are highly charged political issues, and it is difficult if not impossible to separate the purely scientific aspects of such measurements from the politics of the situation. Global warming has some things in common with the measurement of the age of the Earth, too, as we shall see.

Let's begin with what is universally agreed upon. First, the "greenhouse effect" on the planet Earth is real. If it were not, then whatever life existed on Earth would likely be vastly different. The basic greenhouse effect is not evil; it is vital to life as we know it. Second, certain gases such as carbon dioxide (CO_2) are effective "greenhouse gases"—that is, they are effective at trapping the heat radiated by the Earth and keeping it from escaping into outer space.

In addition to the two points above universally agreed upon, there is another important point that is nearly universally agreed upon as well. That is that the level of greenhouse gases such as CO_2 is increasing in the Earth's atmosphere. It is also generally accepted that most of the increases in these gases in the atmosphere are due to industrial emissions (the burning of fossil fuels).

Finally, few deny that the average temperature of "the Earth" depending on where and how it is measured, is increasing. Temperature trends over the past 50 years are between 0.5 and 1.0°C warmer than the average over the past 1,000 years.

A somewhat more controversial conclusion, based on all of the above, is that CO_2 and other greenhouses gases, emitted by cars, power plants, and other industrial sources, are responsible for a significant amount of the observed increase in temperatures in the Earth's atmosphere. The list of scientists and organizations who agree with this last conclusion is long indeed. The list of those experts who disagree to some extent with this conclusion, while probably shorter, is important also.

Two main controversies flow out of the above: First, what does all this mean? If indeed humankind is accelerating the warming of the planet, what might this cause? Are we to expect huge increases in severe weather (droughts, hurricanes, blizzards, and so on), will there be massive melting of polar ice and precipitous increases in sea level? Some experts think so, while others demur. Second, and most difficult of all, what do we need to do about it? Need we take no action, or should we, at great cost, drastically cut back on our use of fossil fuels?

The Greenhouse Effect

The "greenhouse effect" in a greenhouse and that of the planet Earth's atmosphere are not the same thing, even though they have the same name. That said, even if you've never been in a greenhouse, you are likely to be familiar with its famous effect. Park your car outside on a bright sunny day and come back a few hours later. The air inside your car (not to mention the dashboard, seats, steering wheel, etc.) will all be warm—hot even. It will be much hotter inside your car than outside. Components for use in the interior of cars are designed by the car companies to withstand temperatures of 180°F (82°C). The inside of a car is a severe environment indeed! This effect is obviously dramatic in the summertime, but even in the winter, when the outside air is cold and the Sun is lower in the sky, there is a significant warming effect inside your car on a sunny day.

The Sun's energy is contained in its rays, which pass through the windows of your car almost unchanged (we say window glass is relatively "transparent" to sunlight). That is, the glass in your car windows does not reflect (cause to bounce off) or absorb much of the energy in the rays of the Sun. The energy in sunlight is significant. At the equator at noon on a cloudless day on the first day of summer, solar energy strikes the Earth at a rate of about 1,000 watts per square meter—about the equivalent of a hand-held electric hair dryer. The figure is less at other times of the day and year and at other locations on the planet or if it's cloudy, but nonetheless a lot of energy is contained in sunlight. Most of that energy passes right through the glass in your car's windows, where it gets absorbed by the stuff that is inside your car—the dashboard and other components, and anything else inside the car. As those things absorb the Sun's energy, they get warmer and warmer. As they warm up, they in turn radiate heat. That radiation tends to occur in the infrared range—these are commonly called "heat waves." That is, the wavelength of radiation from the hot components inside your car consists of (relatively) long infrared rays. Relatively long, that is, compared to the shorter wavelengths of the visible and ultraviolet rays of the sunlight. The longer infrared rays being radiated by the things inside the car are effectively blocked by the glass. Unlike the Sun's rays, they are reflected or absorbed by the glass, and that energy thus remains inside the car, which heats up.

Thus, the (shorter wavelength) energy in the Sun can get through the window glass and into the car, but most of the longer wavelength energy emitted by the hot items inside the car cannot escape back through the windows. The heat is trapped in the car and the inside of the car continues to heat up.

This is a rough analogy to what happens on a much larger scale to the Earth as a whole—except that instead of glass windows, we have a layer of gases we call "the atmosphere." Sunlight passes through the atmosphere and warms the oceans, other bodies of water, rocks, soil, plants, animals, and everything else on Earth. Infrared energy radiated from those things (we call it "heat") rises back up, but much of it cannot escape, because it is absorbed by the greenhouse gases (CO_2, methane, etc.) and water vapor in the atmosphere.

This is where the greenhouse analogy is not especially accurate. The glass walls and roof of a greenhouse do indeed reflect heat waves, but mostly they provide a barrier to keep warm air from escaping—that's why a greenhouse (or your car) stays warm. The way in which greenhouse gases absorb energy in our atmosphere is a little more subtle. Think of a microwave oven. It works by bombarding your food, for example, a potato, with "microwaves," which are a form of radiation (just like sunlight, infrared waves, x-rays, or radio waves). All of these are examples of electromagnetic radiation—where they differ is in their wavelength. The shorter the wavelength, the more intense the radiation. X-rays are very intense and thus have very short wavelengths (less than a billionth of a meter), while radio waves have very long ones (thousands of meters). Microwaves are in between, with wavelengths in the millimeter or centimeter range. It turns out that microwave radiation is "absorbed" readily by water molecules—it has to do with the wavelength of the radiation and the nature of the H_2O molecule. So, when your oven blasts the potato with microwaves, the water inside the potato absorbs the energy in the microwaves (making the water molecules vibrate a lot), and the potato gets hot. This is why dry foods, like crackers, generally do not heat up much inside a microwave oven. There's nothing inside the cracker to absorb all that microwave energy—the waves pass right through the cracker nearly unchanged. Infrared energy—heat waves—have wavelengths a little shorter than microwaves. Infrared waves are to a large degree also absorbed by water and by greenhouse gases such as CO_2 and methane. Thus those gases in our atmosphere heat up when struck by infrared energy leaving the planet, much as a potato heats up when struck by microwaves.

You can do an experiment that dramatically demonstrates the effect of an atmospheric greenhouse gas like CO_2. First, build a clear plastic box. On one side, mount an infrared camera. (This is the sort of thing a soldier uses in his "night vision" goggles. That way, the soldier can see people or vehicles—anything that gives off heat waves—even at night.) The infrared camera will show a clear image of, say, a person sitting on the other side of the box—assuming the box is filled with air. Now comes the interesting part. Slowly inject CO_2 gas into the box. As the level of CO_2 (an invisible gas) in the box rises, the image of the person on the infrared camera will gradually disappear, because the CO_2 is absorbing the heat waves coming off the person, and thus those waves never get to the camera, and the camera can no longer "see" the person.

Since so much of the Sun's energy is absorbed by greenhouse gases and cannot escape, the Earth stays warm. One can estimate through calculation how much colder the Earth would be if it did not have an atmosphere of heat-trapping (greenhouse) gases. (The main gases in our atmosphere, oxygen and nitrogen, are "transparent" to infrared waves. They do not absorb infrared energy and thus do not trap any of the heat escaping from the Earth.) The first person to calculate how much colder the Earth would be without greenhouse gases was the French scientist Joseph Fourier in the early 1800s. Fourier might thus be thought of as the father of modern-day global climate modeling. Modern versions of Fourier's calculations allow us to conclude that the planet would be about 59°F (33°C) colder on average than it is now. This is a huge difference—easily enough to ensure that whatever life that could exist on the planet would be substantially different from what we have today.

It is thus a good thing that the Earth stays warm, aided by its insulating layer of greenhouse gases. But how much of a good thing is too much? A glass of red wine with dinner is said to be good for one's health. But that doesn't mean you should suck back a full bottle for breakfast. CO_2 levels on Earth have gone from about 315 parts per million in 1958 to over 370 parts per million today—an increase of nearly 20%. The famous "Keeling curve," named for the late Dr. Charles Keeling and based on his decades-long work

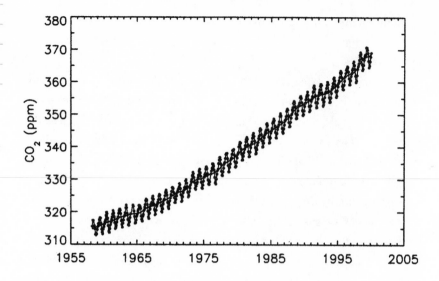

Fig. 8.1. The Keeling Curve of atmospheric carbon dioxide from Mauna Loa, Hawaii, showing the broad trend in increasing levels of atmospheric carbon dioxide. Courtesy of the National Oceanic and Atmospheric Administration.

in Hawaii, is shown in Figure 8.1. The broad trend shown by these data is one of ever-increasing levels of CO_2. Superimposed on that broad trend can be seen the seasonal effects (the saw teeth in the curve—one tooth per year), as plants absorb carbon from the air during the summer growth season and release it during winter decay.

Carbon In Equals Carbon Out

The combustion of carbon-based fuel (gasoline, diesel fuel, jet fuel, heating oil, natural gas, coal, etc.) creates huge amounts of CO_2. "Carbon in equals carbon out" is the way things are in an internal combustion engine. That is, the carbon that flows into an engine in the form of, say, gasoline, has to come out the tail pipe of that engine in some form. The most likely (and, truth be told, the best) form for that carbon to take is CO_2. Despite what some people think, CO_2 is not poisonous.

Alternative forms the carbon might end up in, such as carbon monoxide (CO) and partially burned hydrocarbons, are resolutely to be avoided. They are universally acknowledged to be disasters to human health. Emissions-control technology has improved enormously over the past thirty years or so, both for automobiles and in the "smokestack" industries, such as electric power generation. Atmospheric levels of the "Big Three" tail pipe pollutants, CO, oxides of nitrogen (NOx), and hydrocarbons, have all been drastically reduced due to widespread use of catalytic converters and computerized engine controls on automobiles and other internal combustion engines. A modern car pollutes less per minute of operation, by almost any measure, than a small gasoline-powered lawnmower. (I can't resist adding that the car is a heck of a lot quieter, too.)

We have effectively demonstrated our ability to control known poisons such as the Big Three. But we cannot change "carbon in equals carbon out." To lower the rate at which we pump CO_2 into the atmosphere, we have only three options (or some combination thereof). First, we can increase the efficiency of carbon-fuel technology (less carbon in equals less carbon out). Second, we can lower the rate of consumption of carbon fuel through public policy or lifestyle changes. Third, we can convert to a non-carbon-based fuel technology, such as the much ballyhooed "hydrogen economy."

Just how much carbon do we put into the atmosphere? If your car gets 18 miles per gallon, you pump about one pound of CO_2 out your tailpipe for every mile you drive. In its solid form, CO_2 is called dry ice. A pound of dry ice would be an "ice cube" about 2.5 inches on a side. Thus, one pound per mile of CO_2 emissions is like depositing a 2.5 inch cube of dry ice on the road for every mile you drive. That is also equivalent to 12,000 pounds a year (if you drive the typical 12,000 miles per year), or about four times the weight of a Honda Accord. If your car gets 36 miles per gallon, you have cut your CO_2 emissions in half, to only half a pound per mile, or 6,000 pounds a year. That's still a lot carbon, especially when we multiply by all the cars on the planet and add in all the other significant man-made sources of carbon. In fact, global carbon emissions (from cars and other sources) increased from about 1.5 billion tons per year in 1950 to about 6.5 billion tons per year in 2000. As noted above, over roughly that same period of time, atmospheric CO_2 concentrations have increased about 20%.

A Cynical View of the Global Warming Debate

There is a rather cynical way to look at the whole carbon-emissions controversy. In making the claim that man-made carbon emissions are dangerously warming up the environment, those who make this charge are playing a sort of "environmental trump card," as noted by Bjorn Lomborg. There is almost no way to significantly reduce carbon emissions without major lifestyle or economic changes. Either we have to switch to a noncarbon, probably hydrogen, economy, or we have to reduce our transportation and other energy-consumption habits drastically. Cynically, one could argue that until now environmentalists have been stymied at every turn in their attempts to turn us all into "tree-huggers." Let's turn a cynical eye on several of the major milestones in the history of the environmental movement.

In the 1960s, Rachel Carson wrote *Silent Spring* and many of us became aware for the first time of the ways in which humankind was poisoning its environment. Some argued for a return to the country and a simpler lifestyle. But society's response, once it awoke to the enormity of the problems, was to pass legislation and develop technologies that allowed us to clean up polluted rivers and lakes and to improve the air we breathe. While few would argue that the issues raised in *Silent Spring* have been completely resolved, the improvements have in many cases been extremely impressive. For example, if one compares the level of tailpipe pollutants (carbon monoxide, oxides of nitrogen, and unburned or partially burned hydrocarbons) produced by one of this model year's automobiles to one produced prior to 1969, the level of improvement is truly amazing. And perhaps the most dangerous tailpipe pollutant of all, lead, has almost completely disappeared, for we no longer add lead to gasoline. A modern car pollutes less, as noted earlier, than a small gasoline-powered lawnmower, which is not yet required to adhere to any pollution control limits. Technology has, in a certain sense, trumped the environmentalists and allowed us not only to be able to keep our profligate lifestyles but to make them even more so. If environmentalists get the feeling that the rest of society is thumbing its nose at them, as we roar past their bicycles in our SUVs, it's easy to see why.

After *Silent Spring*, there were the energy crises of the 1970s. Energy prices spiked and there were severe spot shortages, especially of gasoline. Baby boomers today tell their (typically bored and unimpressed) children stories of waiting in line for hours to fill the family car with gas. Once again, environmentalists warned that only huge increases in energy efficiency coupled with unpleasant, drastic changes in lifestyle could save us— otherwise we would soon run out of energy altogether. But technology triumphed again. We did increase industrial efficiency—American gross domestic product (in constant dollars) per unit of energy consumed has increased about 50% since 1970. At the same time, we've discovered vast new energy resources (typically hydrocarbons) with which to stave off the doomsday predictions. As a society, our lifestyles did not have to change. On the contrary, in the United States, we drive ever bigger cars and trucks over much longer distances annually (total vehicle-miles per year have more than doubled since 1970), nearly every one of our cavernous new houses is air-conditioned (even way up north), and our lives are filled with an amazing array of electronic gadgets that were the stuff of science fiction back in the 1970s. The environmentalists were trumped again.

To conclude the cynical argument, those pesky environmentalists are after us once more—this time over global warming. It seems they will not stop until they have forced us to change our lifestyles. And this time, they might have us over a barrel. If the environmentalists succeed in convincing the world that global warming is (a) real, (b) going to get worse, (c) caused (at least significantly) by people, and (d) bound to wreak havoc on our environment, we shall have no choice but to change from a carbon-based energy economy and institute profound changes in our lifestyles. I wouldn't bet on any of that happening, but I'm not a betting man.

Although I have presented at some length a cynical argument for the environmentalist position on the global warming debate, I prefer to be a little more optimistic. Environmentalism has had an overwhelmingly positive impact on our environment. Things may not have turned out the way many environmentalists would have liked (in their worst nightmares, they couldn't have anticipated the ubiquity of SUVs and McMansions). But the environmental movement can take pride in the reduction of air and water pollution, improvements in energy efficiency, increases in public awareness, and so on, that have taken place since the 1960s. The same carmakers who had to be dragged kicking and screaming into creating modern emissions-control systems, for example, now beam with pride at the fruits of their labors.

This is all intended to frame the global warming debate in terms similar to that of the question of the age of the Earth. Lord Kelvin measured the age of the Earth largely because he wanted to discredit the uniformitarian geologists. This does not mean that he fudged his numbers or "cheated" in any way. He was a true scientist, but one who, like all of us, had to live in his time. Today, many of those developing the science related to global warming must certainly take a great personal interest in the debate and hope that the scientific results, and the resulting policy changes, are in line with their own personal views. Let's hope that they are able to put aside those feelings and be as dispassionate as possible. Just to make the required measurements, and to interpret them, is difficult enough.

Asking the Right Questions

One major difference between global warming and the age of the Earth is the basic questions themselves. As Arthur Holmes put it in 1913, "it is perhaps a little indelicate to ask our Mother Earth her age." Indelicate, maybe, but rather straightforward. What about the global warming question, or should I say questions? Most of us, I imagine, have a vague idea that there are a lot of serious-minded (and well-meaning, although probably from the left of the political spectrum) scientists who truly believe that the Earth is getting warmer and that "technology" is somehow to blame. Most have likewise heard the term "greenhouse gas," but the percentage of us who could accurately define what that means and give examples is probably pretty small. This is not to belittle anyone or anything (such as science education in the United States). It's just that there is a lot of complicated science going on here, in a wide variety of scientific disciplines. For a long while in the early days of this story, the scientists themselves did not want to believe what their measurements were telling them.

The age of the Earth has one overall measurement associated with it—the age of the Earth. There really isn't a similar single number associated with global warming. Instead, a whole suite of measurements vies for our attention. There are the data on temperature deviations from the average over some period. Among the best-known representations of these data is a graph that has come to be known as the "hockey stick" (Fig. 8.2), which shows deviations from the average temperature over the past millennium. From the year 1000 to 1900, the graph, while exhibiting short-term fluctuations in a manner reminiscent of the stock market, shows a broad trend that is slightly but steadily downward. The years 1000–1900 thus represent the handle of the hockey stick. For the years 1900–2000, by contrast, the temperature deviation graph curves up abruptly—representing the blade of the

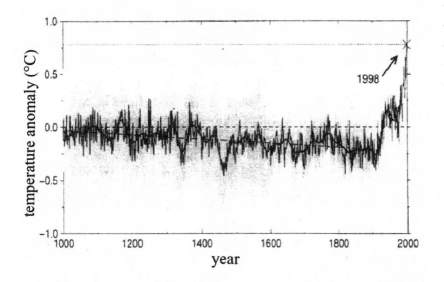

Fig. 8.2. The famous "hockey stick" graph, showing a compilation of actual measured temperatures (since 1902) and temperatures reconstructed from other measurements (since 1000), presented as deviations relative to the average temperatures in the Northern Hemisphere from 1960 to 1990. Reproduced by permission of the American Geophysical Union.

hockey stick. The rise is particularly sharp for the 50 years from 1950 to 2000. The hockey-stick shape of the graph is dramatic. You may find the magnitudes of the actual temperature deviations to be less so. Even for recent years, such deviations are only about 0.6–0.7°C (1.1–1.3°F) from the overall average over the past millennium. Even so, the hockey-stick graph has come to serve as a sort of overall measurement for activists in the global warming movement. Taken as a whole, the science related to global warming is overwhelming, but anyone can relate to the hockey stick.

The hockey-stick graph was first published in 1998 by Michael Mann of the University of Virginia and his associates. Because of its dramatic shape, it quickly became a symbol for the entire global warming movement—the graph seemed to shout out all at once, "It's getting warmer, and it's our fault!" In the past few years, however, a variety of scientists, including a number of statisticians, have found significant errors with both the data that went into the hockey stick and the means by which those data were analyzed.

Those data are voluminous and come from a variety of sources. Creating a graph like this, which purports to show fluctuations in our planet's average temperature over a thousand years, is a big job. Building such a graph reliably is beyond the ability of all but a very few specialists. Even understanding it, on anything but a superficial level, is difficult enough.

The debate over the hockey stick in the past few years has become rather heated. The question, "What should we do, if anything, about global climate change?" threatens to become instead, "Do you believe in the hockey stick or not?" It's worth pointing out that even some experts who think the hockey stick is "rubbish" still believe global warming is real and worthy of action.

When it comes to the weather, most of us are pretty selfish. Is it going to rain tomorrow and ruin my tennis game? Am I going to freeze when I visit Winnipeg in November? And why does it always have to be so humid in Houston? These are the types of weather-related issues that I tend to get worked up about. Frankly, why should I care that the Earth may be, on average, about 1°F warmer over the past few decades than it was for the previous 1,000 years? Even when we hear that the polar ice caps may be melting and that sea levels may thus rise by several feet over the next century, many of us can't get too excited about it.

Bjorn Lomborg's book *The Skeptical Environmentalist: Measuring the Real State of the World* (2001) is packed with charts and graphs related to human health, hunger and prosperity, land use, energy production and use, water and air problems and resources, waste disposal, environmental toxins, and global warming. Lomborg, by his own admission "an old left-wing Greenpeace member," is also a professor of statistics—and we professors can be fickle. He became convinced that the average educated person's views on environmental issues are based not on the actual measurements that exist, but on doomsday predictions in the popular media—where dire predictions sell a lot better than sober, balanced analyses. He decided to cut through the hype on environmental issues and take a cold, calculated look at the numbers. He found that a lot of the things he believed in as an environmentalist—bad things about the environment—simply were not true. As a result of looking at the numbers, he had a lot of rethinking to do. Because of his environmentalist leanings, and not in spite of them, *The Skeptical Environmentalist* is an all the more impressive effort.

Lomborg found there are lots of numbers to look at. Depending on the environmental issue considered, Lomborg's conclusions vary (unlike *The First Measured Century*, *The Skeptical Environmentalist* is not just data— Lomborg interprets the data and is not afraid to make his opinions known). However, in general, Lomborg concludes that his original premise is mostly (but by no means entirely) valid: lots of the doomsday predictions about the environment are not supported by the measurements. In many areas of the environment, he says, things are not nearly as bad as we've been made to believe by the popular media. For example, he concludes that we are not running out of energy any time soon (or water or food, either) and that we have made great strides in improving air and water quality. I don't want to either oversimplify or sugarcoat his analyses: he concludes that along with the good news, there remain serious challenges in each of these areas as well.

In terms of global warming, the story is much the same. Lomborg concludes that there are significant challenges, but that most of the hype (the predictions of drastic increases in severe weather, catastrophic increases in sea level, unprecedented droughts, and so on) is not warranted *based on the data and on the best available predictions*. No one can predict the future with certainty. Lomborg bases his conclusions on what has happened (what we have measured) and on how we can use those measurements to predict (through computer models) what will happen.

Not surprisingly, Lomborg's work has plenty of critics. The environment is an extremely emotional topic for many people—and why shouldn't it be? As a result, the Lomborg debate sometimes veers away from reasoned scientific give-and-take and descends into shouting and name-calling. When it doesn't, when the debate over Lomborg's work remains reasoned and quantitative, it is instructive and serves, I believe, to reinforce Lomborg's main thesis—that too much of the environmental debate in general is driven by hype and doomsday prophecies in the popular media, and not on careful consideration of the available measurements.

One cannot hope to get a grasp on the global warming debate without considering the "crystal ball" aspect of the problem. Global climate models (or GCMs) are crucial. Since there is only one Earth, and we live on it, we cannot really do much in the way of experiments to predict global warming.

To investigate global climate change, we can really only do three things.

First, we can measure what's going on right now. We can, for example, measure temperatures on Earth, in the oceans, and in the atmosphere; we can measure the level of CO_2 and other gases and aerosols in the air; and we can measure wind velocities and atmospheric pressures.

Second, we can make other measurements and, from them, infer how things were in the past. Even though we have no direct temperature measurements (systematic atmospheric temperature measurements only go back a century and a half; before that there are sporadic results, such as the diaries of Thomas Jefferson, and there's nothing at all before about 1650), we can get a good idea how warm it was a thousand years ago by inference—by examining, for example, the growth rings of trees or the compositions of gases trapped in layers of ice miles below the surface of Greenland or in Antarctica. Thus, we can measure what it's like now and we can measure what it was like in the past. Third, there is the future, which we can only try to predict. Predicting the climatic future involves ever more sophisticated computer models.

Global Climate Modeling

Attempts to model the climate mathematically go back to about 1870, but the early pencil-and-paper efforts, although quite elegant, were rather simplistic. The early modelers were forced to leave out really important factors—ones they knew were crucial—because including those factors would have made the required calculations practically impossible. There were way too many numbers to crunch. Even with today's computers, and with several decades of GCMs to build on, the best climate models are far from perfect.

The climate is so complex, it's hard to imagine how a computer program could accurately predict its future. It can't—at least not yet. All you have to do is watch the weather forecast on TV to see that. Computerized weather prediction models do reasonably well at predicting weather over the next 48 hours, but if you go much past that, well, we've all seen the results. But weather prediction models aren't the same—they don't have the same goals—as global climate models. I want a weather model to tell me if it's going to rain in my town on Saturday, or if the hurricane that's gathering strength in the middle Atlantic is going to veer south toward Florida or north toward Virginia. A global climate model, on the other hand, is expected to predict how warm the planet will be, on average, 50 years from now, or 100.

So many important factors go into a global climate model, it's hard even to name them all, let alone to take them all into account mathematically. These models must predict the influence of wind, humidity, clouds, the Sun, the oceans, the various gases in the air (such as CO_2), particulates in the air (the so-called aerosol effect), changes on land (such as the amount of rain forest), and so on. To make matters worse, many of these effects interact with one another, and the models have to be that much more sophisticated to take these interactions into account.

We have had global climate models for some time now—long enough that we can compare their predictions from a few years ago to actual data that have since been gathered. One problem early in this process (back in the 1980s and early 1990s) was that increases in CO_2 tended to overpredict increases in average global temperature. In other words, the models tended to be too sensitive to CO_2. Gradually, climate modelers were beginning to realize that other factors were at work. The role of aerosol particles was just beginning to be understood and accounted for in the models. Aerosols include things like dust, chemical haze, and the soot from both industrial and natural sources, like volcanoes. Aerosols can have a cooling effect (they can block solar radiation), but they can also trap radiation and prevent it from leaving the Earth's atmosphere (a warming effect), and they have complex interactions with clouds. Taking aerosols into account was a big leap in sophistication for the modelers.

When Mount Pinatubo in the Philippines erupted in 1991, global modelers had a chance to check their work against a rather large set of actual data. The volcano created a cloud the size of Iowa, containing twenty million tons of sulfate aerosols (roughly the weight of thirteen million full-sized cars). In the wake of the eruption, climate modelers cranked up their best models and predicted that global temperatures would decrease on average by about a half a degree Fahrenheit, that the temperature decreases would be concentrated in the northern latitudes, and that the effects would last a few years. Amazingly, they were pretty much right on with these predictions. This had several important results. First, global climate modeling was thereafter viewed more favorably by many of the skeptics in the scientific community. Second, the role of CO_2 was seen in a different light.

With the sometimes mitigating effects of aerosols now better accounted for, global temperature increases were being much more accurately modeled, as revealed by comparisons with actual measurements.

What About the Ice Age?

Determining the age of the Earth was a great intellectual challenge to geniuses like Kelvin and others. A similar intellectual challenge has existed since those days as well—one that provides some links between the age of the Earth and the global warming debate. I'm referring to the cause of ice ages on Earth.

Today, every schoolchild knows that the Earth has gone through a series of so-called ice ages, during each of which the Earth's climate was cold enough that enormous glaciers covered large areas of the planet and extended well south into what is now the continental United States. Before the early 1800s, most mainstream scientists did not believe that there had ever been ice ages. Those who held with uniformitarian theory generally did not even believe that the glaciers that did exist (it was hard to deny the existence of glaciers) had ever been very different from the way they were then. I guess they figured that if there was a giant sheet of ice over there on the side of that mountain, it had probably been there as long as the mountain itself—"forever." Eventually, it was established beyond doubt that the glaciers had not once but many times been massive and far-reaching, and that long stretches (many thousands of years) of cold weather like this seemed to come and go on Earth in a cyclic fashion. The frequency of ice ages—how often they came and went—was stubbornly difficult to measure. A further challenge became to explain why. Why does the Earth go through periodic ice ages?

Attempts at measuring the periodicity of the ice ages were even related to the age-of-the-Earth question. By figuring out how often ice ages had occurred, and by estimating how many of them there had been, one could perhaps gain some insight into how old the Earth was. A good many explanations for the ice ages have been proposed. These include geological explanations: massive earthquakes that could reshape mountains and change wind patterns or ocean currents could bring on extreme climate change. Then there are the oceanographic explanations: so much energy is contained in ocean water, and the prevailing currents distribute that energy throughout the planet. Small changes in the temperature or salinity of the water could destroy the equilibrium, redistribute energy (warmth) throughout the planet, and change the climate drastically. What about biology? Some have argued that massive changes in vegetation (as a result of either natural or human causes) could shift wind and weather patterns and eventually plunge the planet into an ice age.

Finally, there are the astronomical arguments for the ice ages. Stars vary in brightness over time, and our Sun is a star. Could not a period of abnormally low energy from the Sun conceivably cause an ice age? It is also known that the inclination of the Earth's axis and the shape of its orbit with respect to the Sun vary slightly in patterns that last tens of thousands of years. Might not these variations result in differences in solar energy reaching the Earth—differences that could result in dramatic climate changes?

Eventually, the last of these astronomical arguments found favor. Milutin Milankovitch, a Serbian engineer working in the 1920s and 1930s, improved on earlier calculations of three aspects of the geometry of the Earth's motion around the Sun. First, there is the variation in the inclination of the Earth's axis with respect to the Earth's orbital plane around the Sun. In other words, the Earth does not stand "straight up and down" as it orbits around the Sun—it is tilted somewhat (this axial tilt is known as "obliquity"). The more tilted the Earth is, the more variation we shall see in seasonal temperatures—more tilt thus means hotter summers and colder winters. Earth's tilt varies between $21.5°$ and $24.5°$ over a cycle of about 40,000 years. We're at about $23.5°$ these days.

The second of Milankovitch's geometric factors is the shape of our orbit around the Sun—the Earth's "orbital eccentricity." During cycles that last about 90,000 years, the Earth's orbit around the Sun varies from an almost perfect circle (in geometric terms, we say nearly zero eccentricity) to a very slightly elliptical shape. If the Earth's orbit is a perfect circle, the distance from the Earth to the Sun is constant. The maximum eccentricity of our orbital ellipse is only 0.07—this is not a lot of "stretch"—but during such an orbit of maximum stretch, this means that at its closest point, the Earth is five million kilometers closer to the Sun (about 3.3% of the Earth–Sun distance) than it is at its farthest point. As with the changes in axial eccentricity, changes in orbital eccentricity can accentuate the differences in the seasons, summer and winter.

The third and final geometric factor that Milankovitch investigated relates to the periodic wobbles that occur (the precession of the equinoxes, to use the technical term) in the Earth's axis. These have been likened to the behavior of a spinning top that is beginning to spin down. As the top slows down, it begins to wobble about its spin axis. A little bit of wobble at first, then a lot. For the Earth, these wobbles can have the effect of decreasing the seasonal contrast on one hemisphere (Northern or Southern), while increasing it in the other.

Milankovitch combined his tedious calculations of the above three geometric variations in the Earth's orbit with a different set of calculations that show how the amount of solar energy (known as insolation) absorbed by the Earth varies with each of the geometric factors. (In this, he was guided by theories first developed by Kelvin.) Milankovitch's stated goal was a mathematical theory of climate.

At first, his calculated predictions for the timing of the ice ages were dismissed, because they did not match the accepted chronology, which was based on geological data. In addition, the changes in sunlight striking the Earth that were predicted by Milankovitch were extremely small, and experts had trouble believing that such tiny changes could plunge the entire planet into the frigid grip of an ice age lasting thousands of years.

Little by little, however, Milankovitch's approach to the problem began to gain favor. Measurements using a variety of techniques from a number of sources seemed to point to "Milankovitch cycles." Ice cores removed from Greenland and sediment cores removed from ocean bottoms, among other things, contained data that showed that average temperatures on Earth correlated well with Milankovitch cycles.

Among the profound implications of this was the realization that the very small changes in insolation during a Milankovitch cycle could indeed plunge the Earth into an ice age. In human terms, an ice age comes on extremely slowly, over hundreds or thousands of years (in geological terms, such an onset is considered rapid). Consumed as we are by day-to-day weather changes, it's important to keep in mind that one or two cold winters in a row is not going to bring glaciers to Georgia during our lifetime. But a small reduction in insolation, as Milankovitch predicted, can result in more snowfall. If less of that snow melts (also a result of the decrease in solar energy) the process can begin to feed on itself. This is because snow cover reflects more sunlight than bare earth or vegetation, and thus the small reduction in sunlight predicted by Milankovitch becomes amplified—the Earth receives less and less energy each year from the Sun. This process, known as "feedback amplification" is well known in a variety of applications from electrical engineering to weather prediction. Over hundreds or thousands of cycles (one year is one cycle), small reductions in sunlight—caused by tiny changes in the motion of our planet around the Sun—become amplified into a full-blown ice age. The end of an ice age is then brought about as the Earth's motion gradually allows more and more sunlight, slowly warming things up and melting the ice. Feedback amplification brings on ice ages "suddenly," in geological terms. Lacking a similar feedback mechanism, ice ages tend to end more slowly.

If, as experts believe, Milankovitch cycles control long-term climate change on Earth, why are so many people worried about the effects of increases in CO_2 in the atmosphere? Just what does the ice age story have to do with global warming? Lots, actually. There are, I believe, three lessons to be drawn from this. The first is that global climate change, in human terms, is slow. As we debate the effects of CO_2 on our climate, and in what ways this should shape public policy, it is important to keep in mind that we are really debating our grandchildren's climate, or their grandchildren's, and not our own. This does not make the debate any less important. It means we have to look beyond what we see and feel as human beings. A hot summer does not mean "global warming is here to stay," any more than a cool one means there is nothing to worry about.

As well, the seeming increase in the number of "monster" hurricanes, such as Katrina in September 2005, is taken by some as a clear manifestation of global warming, since ocean temperatures, which do appear to be rising, are a key factor in hurricane strength. But one monster hurricane, or even a season's worth, do not "prove" global warming, any more than a relatively quiet hurricane season would disprove it. Hot summers, warm oceans, and monster hurricanes, taken together and among other evidence, are looking more and more persuasive, however. Were this a jury trial, it seems likely that "a preponderance of the evidence" would be found to support the global warming thesis.

The second lesson I take from the story of the ice ages is the one related to feedback amplification. Milankovitch showed that tiny reductions in insolation could feed on themselves and cause dramatic climate change. Many experts in the global warming debate caution that continued increases in CO_2 could trigger a similar feedback amplification. One feedback scenario holds that more CO_2 in the atmosphere causes a little bit of warming, which allows the air to contain more moisture, which further increases the greenhouse effect, and so on. Once again, in human terms the changes are slow (see lesson one above).

The third lesson from this story is related to the second. This is the "delicate balance" lesson. Many experts believe that our climate is a "capricious beast," and that by increasing the level of CO_2 in it, we are "poking it with a sharp stick," as Spencer Weart puts it in *The Discovery of Global Warming* (2003). We have come a long way from the uniformitarian view that things will always be the way they have always been. Such is our dominion over the planet that we may even be in danger of unwittingly changing the climate for future generations. The CO_2 that we pump into the atmosphere could be the sharp stick—the catalyst for changes—the results of which not even the most sophisticated global climate models could predict. Those who hold this view say: Why take a chance—why risk upsetting the delicate balance that is our climate?

A Fool for a Cigarette

Back when my parents' generation was young, in the 1930s, 1940s, and 1950s, many more Americans smoked cigarettes than today. (*The First Measured Century* tells us the percentage of American men who smoked went from 59% in 1955 to 28% in 1997, while women smokers decreased from 31% to 22% over the same period.) My parents both quit around the time of the first Surgeon General's health warning against smoking in 1964. Before the Surgeon General's warning, that is, before the first official government pronouncement that smoking was bad for your health, it was relatively common knowledge that cigarettes probably were not all that good for you. They weren't called "coffin nails" for nothing. What did parents tell kids back then to try to keep them from smoking? "It will stunt your growth," or "It will turn your teeth yellow," or "It gives you bad breath," according to my mother, were among the admonitions she received as a young smoker from her own mother (who smoked as well).

Nowadays, we all know why we shouldn't smoke. We risk lung cancer, other forms of cancer, emphysema, and heart disease. Not today or tomorrow, but many years down the road. Smoking can cut our life short by a decade or more. Might we thus say that smoking cigarettes upsets the "delicate balance" that exists in our lungs and could lead to "long-term change" when lung cells mutate into cancerous forms, for example?

How apt is the smoking analogy to global warming? It's certainly not perfect, but I think it's worth considering nonetheless. Many people smoke for 40 years or more and die of unrelated causes. Smoking evidently does not upset everyone's delicate balance the same way. What will happen to the delicate balance of our climate as the level of CO_2 in the atmosphere continues to rise? No one can say for sure—we simply cannot predict the future.

Bjorn Lomborg has looked at all the numbers. He believes it is likely that our economy by the mid twenty-first century will largely have shifted (due to economic factors and improved technology) away from one based on fossil fuels to one based on hydrogen, solar energy, and other renewables. He further believes that the numbers show that this will reduce the rate of increase of CO_2 in the atmosphere before CO_2 levels will be able to upset the delicate balance. Lomborg thus urges caution as we consider public policy changes to combat global warming. Some models show that the costs of drastic intervention could even exceed those of doing nothing. (Models that predict global climate change by combining traditional weather factors with economic and population growth factors are now common.)

It is almost impossible to consider "what to do about global warming" without plunging into the world of computer modeling. Such a plunge can be bewildering, given the sophistication of the models and their conflicting conclusions. This is all part of the measurement story, as we shall see in the next chapter.

CHAPTER 9

GARBAGE

IN,

GARBAGE

OUT

The Computer and Measurement

*Any sufficiently advanced technology is
indistinguishable from magic.*
—Arthur C. Clarke, *Profiles of the Future* (1973)

Google That

When I type "measurement" and "computer" into the search line at
google.com and hit the return key, I see displayed almost instantly on my
screen the names of the first 10 web sites out of "about 5,110,000" that are, to
varying (and measurable) degrees, relevant to my request. If I wanted to, I
could peruse all 5,110,000 of those sites (or I could first refine my request—
even professors don't have that kind of free time). The technical term for
what Google has done for me is a "miracle." Well, actually that's not the
technical term, but that's what it seems like to me—a guy who is supposed to
know something about technology.

Whatever you call it, modern Internet searching involves a lot of measurement. Google measures the "relevance" of the web pages it provides me with using proprietary software. One of the factors that goes into measuring relevance is the number of links to a given web page from other pages. Each such link is a "vote" for the page being linked to, and some votes are given more weight in the relevance calculation than others—depending on how important the page is that is casting the vote. These web page rankings are combined with a user's search query, such as mine for "measurement" and "computer." Google matches my search words to text on the web pages. In doing so, it claims to go "far beyond the number of times a term appears on a page and examines all aspects of the page's content (and the content of the pages linking to it)." The goodness of the match between the query and a given web page is then combined with that page's rank to determine where it will fall on the list of matches supplied in response to any given query.

Online searching is certainly not the only intersection between the computer and the measurement revolution. The ubiquity of the computer in modern life is hardly newsworthy these days. Computers sit on every workplace desk and fill up the classrooms of most every school. Most of us are likewise aware that computers control a great many functions of many modern machines. A typical car in 2005 is said to contain more computing power than all the computers in the world contained in 1950, and a typical laptop these days is said to be more powerful than the on-board computers used during the Apollo moon missions. Computers (microprocessors is a more precise term) also control telephones, toasters, coffee machines, and a host of other modern conveniences. Statistics and lists like these get a little tiresome, as do the "remember when?" stories of baby boomers like me who did time using punch cards and slide rules. Before I stroll too far down Memory Lane, let's focus in on the computer and measurement.

Three Roles for the Computer in Measurement

The role of computers in the measurement revolution is perhaps under-appreciated. Computers are phenomenal measuring machines. Strictly speaking, the computer itself does not do the measuring. That is generally done by instruments such as thermocouples (for measuring temperature), force transducers (for measuring force), and a whole universe of other instruments for measuring the frequency and intensity of light and other forms of radiation, the strength and orientation of magnetic fields, and so on. While the computer itself does not make these measurements, it does play three crucial roles: data acquisition, data reduction or mathematical manipulation, and numerical simulation.

Data Acquisition

A computer is extremely good at recording measurements. What a "recording secretary" it is! Working in tandem with the appropriate instruments, a typical desktop computer can record measurements of pressure, temperature, force, and many other quantities tens of thousands of times per second—and faster. This sounds impressive, but it may not be immediately obvious why it is so revolutionary.

Things were not always this way. The clipboard, once the constant companion of the data taker, has practically disappeared. Getting the numbers used to be the hard part. Now, once one has the right instrumentation, and the right "data acquisition system" (or DAQ—we engineers love a good acronym) with a computer at its heart, getting the numbers is easy. The hard parts are knowing what numbers to go after in the first place, making the best use of all your fancy DAQ equipment to get the numbers, verifying that the numbers you got are good, massaging the numbers into a useful form (as described below), and, last but certainly not least, interpreting those numbers. Those things are all hard enough—without the old-fashioned problem of recording the numbers. That problem has been erased by the computer and associated DAQ equipment. There is an entire, vital industry these days just to do DAQ. Sophisticated high-tech companies like National Instruments have grown up playing this game.

Computers have aided data acquisition in some widely different and unexpected ways. As one example, there is the eBird program of the Cornell University Lab of Ornithology and the National Audubon Society. Backyard bird-watching goes digital with eBird. The members of a network of over 7,000 "citizen scientists" submit electronic checklists of the species they have sighted, the numbers of birds, the dates the birds were sighted, and so on. All of these data are collected, reduced, and then applied in various ways (this is the DAQ part of it). I find the graphic displays of these data fascinating, and I'm not a bird-watcher. You can, for example, watch the northward spring migration of a bird species displayed on a map of North America in a series of week-by-week "snapshots." In early February, the first yellow warblers are spotted in southern California. Florida, south Texas, and Arizona soon follow. By the end of April, the upper Midwest is full of these lovely yellow critters. Other than just being really cool to look at, this has all kinds of applications for specialists, from basic ornithological science to environmental studies. Like a high-tech canary in a mineshaft, the shifting population patterns of birds provide insights into environmental issues such as air and water quality and land use. Ornithologists have always been able to estimate bird populations, migration patterns, and so on, but a program like eBird makes it so much more immediate and thoroughly detailed. Just as we can see things with a high-speed camera that are impossible to discern with the unaided eye, eBird allows us to see things about birds that we could not see before. I suspect that before long there will be many other applications of this unusual form of computerized data acquisition.

Global positioning systems, certainly a vital part of the measurement revolution, are a great example of a technology that relies on computerized data acquisition. Most of us are aware of what a GPS does—especially since more and more cars are equipped with such systems. GPS allows us to track our location (as we drive across an unfamiliar town, for example). Combined with the latest electronic maps, GPS software will find an address for you, compute the best driving route, and even prompt you when it's time to turn.

GPS was developed by the U.S. military for reasons it doesn't take a rocket scientist to divine. The system consists of an array of satellites in geosynchronous orbit above the earth. (This means the satellites are located in space such that their positions relative to the rotating Earth do not change over time. Imagine these satellites as being anchored to the Earth by extremely long, perfectly rigid poles.) The satellites send off signals at precisely timed intervals. The GPS unit in your car receives signals from multiple satellites. This is data acquisition, and your car's GPS contains a DAQ system to handle it. Knowing the time that any given signal was sent by one of the satellites, the time it was received at your car, and the speed at which the signal traveled, the computer in your GPS can calculate its distance from that satellite to within a few meters. By combining the distance measurements obtained in this manner from several different satellites, your GPS unit can calculate your precise location, again to within a

Fig. 9.1. Two-dimensional "GPS" example. The three circles intersect at Tulsa, the only place that is exactly 551 miles from Denver, 359 miles from Saint Louis, and 234 miles from Dallas.

few meters. The math involved isn't all that complicated, but the whole enterprise requires a computer, because everything (the data acquisition and the location calculations) gets repeated over and over again so rapidly.

As a two-dimensional example, let's say you wanted to find your location, but all you know is that you are 551 miles from Denver, 359 miles from Saint Louis, and 234 miles from Dallas. (These three cities, then, represent the locations of the GPS "satellites" in this two-dimensional example.) To find your location, you can draw three circles on a map, each centered on one of the satellite cities. The first circle will have its center at Denver and a radius of 551 miles. The second will be centered in Saint Louis with a radius of 359 miles, and the third will be centered in Dallas with a radius of 234 miles. You will see that the three circles intersect at only one point—my hometown of Tulsa, Oklahoma. Tulsa is the only place you can be that satisfies all the given conditions. Instead of doing this graphically, by drawing circles, you could get the same result mathematically, using equations, which is how your GPS computer does it (and in three dimensions, not just two, as with compass and pencil).

GPS is thus a good example of how computers have revolutionized measurement by revolutionizing both the acquisition and transformation of data.

Data Transformation

The second crucial role that the computer plays in measurement thus involves the transformation of the raw measurements into useful numbers. GPS transforms the actual raw measurements (the time it takes for the signal to travel from satellite to GPS unit) into something much more useful: your exact location. Needless to say, this sort of transformation of data happens in measurement all the time, irrespective of whether a computer is involved. In Chapter 2, we saw one example: the transformation of the diameter of a circle into the circle's area using the simple geometric relationship $A = \pi r^2$, where A is the area of the circle, and r is its radius (one-half the diameter).

There are countless other examples. Think of how an old-fashioned fish scale works. You catch a big fish and hang it from the scale on a hook. Inside the scale is a spring. The heavier the fish, the farther the spring stretches—this is indicated by a pointer on the side of the scale. What you are really measuring, the "raw measurement," is how far the spring extends—the distance in inches or centimeters. But the numbers inscribed along the length of the scale are in pounds or kilograms. It turns out that if the spring is made correctly, it will have a constant rate of extension. This means that a spring might extend (for example) 1 inch for every 10 pounds of force pulling on it. If the fish weighs 5 pounds, the spring stretches ½ inch. If the fish weighs 20 pounds, the spring stretches 2 inches. This demonstrates Hooke's Law: "Ut tensio sic vis" (As the extension, so the force). (Discovering laws of nature evidently wasn't enough for seventeenth-century Englishmen like Robert Hooke and Isaac Newton: they also had to write them out in Latin.) In modern English, Hooke's Law says that the force on a spring equals the extension times a "spring constant." For the spring in our example, the spring constant is 10 pounds per inch.

The numbers inscribed on the side of the fish scale are a computer of sorts—a primitive computer that has a single specialized function: to convert distance in inches to force in pounds according to Hooke's Law. This kind of graphical calculator used to be common, and it still is. Using a pencil and compass to find a location on a map is a great example of graphical calculation. A thermometer uses a similar principle to relate the expansion of mercury or another liquid to its temperature. An old-fashioned mechanical automobile speedometer works in a similar fashion: it converts the rate of rotation of a certain gear in your car's transmission to a road speed in miles per hour using a graphical calculation on the face of your car's speedometer. (That's why, if you are stuck on ice spinning your wheels, the speedometer tells you that you are moving even though you aren't. What you are really measuring is not your road speed, but the speed at which the gears in your car's transmission are rotating. These two measurements are not the same when your wheels are slipping with respect to the road.)

None of the examples of measurement instruments cited above, the spring-type fish scale, the mercury thermometer, or the mechanical car speedometer is particularly amenable to computer data acquisition. All of them could be adapted to the computer (and probably have been), but each of these examples also serves to show how measurement has changed, even for such everyday measurements as these. The spring-type fish scale, the mercury thermometer, and the mechanical speedometer are all examples of analogue (as opposed to digital) instruments. You can buy a digital fish scale or a digital thermometer, and digital car speedometers are found on more and more cars these days—but they all use different technology from their mechanical analogues. One of the ways in which these instruments differ from their old-fashioned predecessors is that the data generated can typically be easily gathered by DAQ equipment. If I catch a whole bunch of fish and weigh them one by one on my digital fish scale, I can download the data to a computer and calculate the average weight and its standard deviation, keep track of the heaviest and the lightest fish, and so on. I can even have the computer e-mail the results to my sure-to-be-jealous fishing buddies back home (along with a digital photo of me living it up in some exotic locale). I could also do all of these things by hand, of course, but it would take far longer.

The simple calculations involved in transforming the raw measurements change, because now our instruments work by different principles. Our digital fish scale contains no spring. Instead, it contains a device called a transducer, in this case a force transducer. A force transducer, when squeezed, gives off an electrical current. (A similar principle is at work in the automatic lighting devices often used on gas ranges and grills. When you turn on the gas, a piezoelectric crystal [the prefix *piezo-* means "pressure"] is repeatedly struck by a little hammer—that's the snapping noise you hear. The sharp force of the hammer is enough to cause the crystal to spark, and this, with any luck, will light the flow of natural gas or propane and you can blacken your redfish or sauté your asparagus.) Getting back to the digital fish scale, when the fish is placed on the scale, the transducer gets squeezed by the weight of the fish and gives off its electrical signal, and a computer or microprocessor then has to convert that signal into "pounds." This is analogous to your eyeballs reading the numbers inscribed on the spring scale to determine what the fish weighs. The transducer has to have been calibrated—so many volts given off by the piezoelectric crystal equals so much force—just as the spring on the old-fashioned scale was calibrated in pounds per inch of extension. Thus the raw data measured by the digital fish scale, volts, is converted to useful numbers—the weight of the fish in pounds.

A Picture's Worth a Million Words

All of this is child's play compared to what computers are capable of when it comes to transforming raw data. Most of the images we see on a computer screen, whether they are photographs taken with a digital camera, ultrasound images of a fetus in the womb, or magnetic resonance images of someone's brain, are the result of the computer mathematically manipulating the results of zillions of individual measurements. In some ways, this is the real measurement revolution. It is hard to overstate the importance of this phenomenon in our lives—and it's becoming more sophisticated and important all the time.

In just a few decades, MRI has become part of everyday health care, as well as a part of our vocabulary. The first image of a human being made using MRI technology was created in 1977. It took five hours and by today's standards was of poor quality. Today, MRI creates beautiful images in seconds and, while not quite as commonplace as x-rays, MRI has become routine. Twist your knee on the tennis court? The doctor is likely to order an MRI as a precaution.

MRI is, in essence, an enormous number of sophisticated measurements that have been manipulated mathematically and repackaged in a form (a color image) that is most useful. An MRI machine (that scary-looking giant doughnut-shaped magnet) goes through a patient's body point by point (a "point" here is a small cube of your body about one-half milli-meter on a side—not much bigger than the period at the end of this sen-tence). Each point is analyzed as to what type of tissue it is—different tissue types respond differently to the radiation inside the MRI machine, and those differences can be measured.

The information on all the points is then integrated into a two- or three-dimensional image, where different types of tissues are shown as different colors. This type of multicolor image is a useful way of displaying measure-ments graphically. Just as a graph of, say the Dow Jones Industrial Average over the past 12 months (Dow Jones on the vertical axis, time on the horizon-tal axis) is a handy way to display the health of the stock market graphically, an MRI is a handy way to display the health of your knee or other body part graphically. But for MRI, this type of imaging is more than just "handy." It's essential. Instead of a graph, I could look at two columns of numbers, one for the Dow Jones and the other for the month of the year, and get a pretty good idea of what has been going on in the market—the graph, though, is much faster and more convenient. I can't do that with the measurements created by an MRI machine (and neither can your doctor, I would venture to say). In other words, the "raw" MRI measurements are not useful. Dis-playing the MRI image is not just a convenient way of representing MRI measurements, it's indispensable to being able to use those measurements.

In that way, the image created by an MRI machine is no different from a digital photograph. Both are multicolor displays of a whole bunch of indi-vidual measurements. A digital camera measures the light intensity (bright-ness) and frequency (color) that are given off, point by point, by whatever it is you are photographing—a flower, say. More expensive digital cameras, those with more megapixels, can divide the image of the flower into more individual points and thus provide a higher resolution image. To each point in the image, the camera assigns a color based on the light intensities and frequencies it has measured, and that color is then displayed on the cam-era's LCD screen, a computer monitor, or, with the aid of a printer, on paper. The image that we see is a useful and convenient way to present millions of light intensity measurements graphically. The image created by an MRI machine differs only in what is being measured. With a digital photo, we're measuring (visible) light intensity and frequency. With MRI, we're measuring the absorption and emission of energy in the radio frequency range with the aid of that giant magnet.

In both cases, MRI and digital photography, it takes a computer to keep track of and process all those measurements (millions of them for a single image). We've been able to measure light intensity (to use the digital photo example) for a long time. (The Greek astronomer Hipparchus's system for measuring star brightness, which dates back to 120 BC, is still used today in modified form.) Dividing up a single image into millions of such measurements requires sophisticated instrumentation, certainly, but above all it requires a computer to organize and keep track of everything—to direct the whole enterprise.

Numerical Simulation

A third major role for the computer in the measurement revolution is in numerical simulation—also known as computer modeling. Using mathematical relationships, sometimes discovered centuries before the development of the digital computer, a computer can model or simulate, often with great accuracy, things that would be extremely difficult if not impossible to measure directly. This is how weather forecasting is done (perhaps not the best example to start with), and it is also how an airplane manufacturer like Boeing or Airbus predicts how well a new airplane design will fly before it ever builds the first prototype.

This isn't "measurement" in the traditional sense. But it is so intimately related to measurement, and so important, that I feel that it needs to be discussed. In many cases, numerical simulations are done instead of more traditional measurements. Take the example of the prototype jetliner mentioned above. It costs a great deal of money to build and test any airplane, much less a passenger airliner. The more "experiments" that you can do numerically, on the computer, without having to actually build the hardware, the more money (and time) you can potentially save. There is also safety to think about. Ask any kid with a flight simulator game on his computer how many times he has crashed the plane. But a video game flight simulator is just a game. Persuading a computer to perform a realistic simulation of a commercial airliner is an immensely complex undertaking, fraught with uncertainties and full of real-life consequences. Not to mention that it costs a lot of money.

Why do we perform such simulations? The answer is simple: numerical simulations are performed when it is cheaper or faster to get the information we need that way than it would be to make the measurements directly. There is an old saying in engineering: "Measure or calculate as long as the cost of not knowing exceeds the cost of finding out."

These numbers, these things that we measure and calculate, have a role to play. They are there to make our lives better, or to improve our understanding of the world around us, or maybe just to satisfy our curiosity. Why should it matter how we get these numbers—through direct measurement or numerical simulation? In the case of measurement, doesn't the end justify the means?

There is in my profession (here I put on my hat as an engineering researcher) a sort of friendly (and sometimes not so friendly) rivalry between two groups: the "modelers" and the "experimentalists." Both are getting more sophisticated all the time (and many researchers dabble in both activities). I am an experimentalist, but I must admit that the modelers are probably getting more sophisticated faster than we are (notwithstanding all the advances in DAQ). When I was in graduate school during the late 1980s, I used to enjoy watching the sparring sessions between two of my professors, one a modeler, the other an experimentalist, both expert at their craft, and both practiced in the art of debate. The modeler was adamant in his belief that the mathematical modeling of complex phenomena was cheaper, faster, and more repeatable (even way back in the 1980s), and that when necessary, the results could be verified experimentally. The experimentalist distrusted the models (not to mention the modelers). He cautioned that too many models were inaccurate, misleading, or just plain wrong. In his opinion, there was nothing like a good old-fashioned experiment when one needed some critical data. This is a debate that continues today (at least in general, and perhaps even between my two former professors as well).

Even when a computer model is mathematically correct, the model is usually highly dependant on the data that are fed into it. "Garbage in, garbage out" is an old saying in the computer world, and it remains true. In the case of numerical simulation, this means that if the initial conditions that you input into a computer model are incorrect, the results predicted by the model will also be incorrect, even if all the mathematical equations in the model are perfect.

A good example is weather forecasting. The data that are inputted into a weather forecasting model, known as "initial conditions," are one reason accurate weather forecasting is so difficult. Tiny changes in initial conditions—temperatures and pressures in the atmosphere—can result in large changes in the results of the model. This is why it is so difficult to predict the path of a hurricane with any precision. We've all seen the satellite images of an enormous hurricane making its ponderous way across the Atlantic Ocean, slowly heading west, on television. But it remains difficult to predict whether an Atlantic hurricane is going to make landfall in North Carolina or Florida a few days later. From the U.S. perspective, Florida had most of the bad luck in 2004, what with Hurricanes Charley, Frances, Ivan, and Jeanne all battering the state. In 2005, it was the Gulf Coast's turn, and the suffering in Alabama, Mississippi, and Texas was exceeded only by the horrific devastation visited upon New Orleans by Hurricane Katrina. Hundreds died, and it will likely be years before that historic city fully recovers. It's worth remembering, however, that hurricane disasters are not limited to the United States. In 2004, Hurricane Jeanne killed an estimated 3,000 people in Haiti.

Florida 2000: Bush Versus Gore

The debate between modeling and direct measurement comes up over and over again—even in some pretty unlikely places. Statistical sampling, a form of modeling, was given quite a serious black eye in the 2000 U.S. presidential election, when exit polls, combined with statistical sampling techniques, were used by the news media to project which candidate, Al Gore or George W. Bush, would win which states.

Statistical sampling is used all the time in the sciences. If you want to know, for example, how many ruby-throated hummingbirds there are in North America, you do not have to go out and count every one of them, which would be impractical, if not impossible, despite eBird. To count the hummingbird population, you rely on sophisticated sampling techniques. If I have a large bag filled with marbles and I want to know how many of them are black, I don't have to count all the marbles. I can get a good estimate by selecting a "statistical sample" of 10, 100, or more marbles and counting the number of black marbles in the sample. That result is then used to estimate the percentage of black marbles in the bag and the "margin of error" in the result, using the mathematical laws of statistics. The larger the sample I choose, the lower the margin of error—it just takes more time and money to do the counting. I can estimate the hummingbird population using similar techniques, by sampling the population in key locations and comparing it to historical data.

For the bag of marbles, the math alone is all we need. It's more complicated for hummingbirds and presidential elections. We can't always get by with mathematics alone. Let's say I want to take a poll to find out who will win the upcoming election for mayor of my town. What I need to do is to ask a small, representative sample of "likely voters" who they'd like to elect mayor. The winner from my sample should be the winner of the election as well. Selecting my sample is not as easy as plucking marbles from a bag, though. I want to make sure the people I select are likely to vote in the mayoral election, since I really don't care about the opinions of those who aren't going to vote. It's not okay simply to select names out of the phone book to call—they may not even be those of registered voters, much less of likely voters. Even the list of all registered voters is not expected to yield a sample of likely voters. Registered voters represent about two-thirds of the population of voting age. But only about 50% of the voting-age public typically votes in a presidential election in the United States. For other elections, the percentage is smaller still.

Pollsters determine whether someone is statistically "likely" to vote through a series of survey questions, such as whether they voted in recent elections, their interest level in various candidates, and their knowledge about the election in question. Someone I might label as likely to vote might very well strike you as unlikely to do so.

The media have been using statistical sampling techniques for some time. They do this both to predict the winner before an election and, through exit polling on election day, so as to be able to announce the winner much faster than if they had to wait for the results of the traditional nose count of the actual ballots. The problem in 2000 was Florida. The vote was too close to call based on the exit polls and sampling techniques, but that did not stop several networks, in their zeal, from declaring first one candidate and then the other the winner. This was the beginning of a comedy of errors that was not settled until the U.S. Supreme Court finally stepped into the fray.

The Florida presidential election of 2000 showed that there can be serious problems not only with statistical sampling, but also with the traditional nose count. To fix any problems with nose counting (such as those infamous hanging chads), we are now trying to introduce computerized voting machines and other technological fixes. Problems reported in the state of Washington and elsewhere during the 2004 U.S. elections do not inspire confidence that the new nose-counting techniques are much more reliable than the old ones. Simply replacing a measurement technology with acknowledged problems (such as punch cards) with computers does not guarantee the situation will be improved. We may merely be replacing hanging chads with other problems, potentially much worse, such as a voting system that can be hacked into by a 12-year-old in Singapore. Throwing computers indiscriminately at the voting problem is not the best way to improve election accuracy, any more than throwing computers indiscriminately at students is the best way to improve education.

Several years before any of us had ever heard of a hanging chad, there was a vigorous and quite entertaining debate in Congress about how the U.S. census should be conducted. The Constitution requires a census to be taken every 10 years. This means you count all the people, one by one, right? That way you know exactly how many people live in each state, city, and congressional district. This is important for all kinds of reasons, not the least of which is the redefining of congressional districts that regularly takes place, state by state, in the United States—a highly political undertaking to say the least.

The U.S. Census Bureau is a venerable institution. In order to do its job better, during the 1990s, the Bureau developed a plan for conducting the census that included the use of computer-based statistical techniques known as "sampling." But to count people, we shouldn't have to do that, right? People can be relied on to respond to mail-in surveys, or, failing that, to door-to-door canvassing, can't they? Not everyone believes that they can. Officials at the Census Bureau, many academic experts, and quite a few elected officials believe that census accuracy will be improved by using statistical sampling techniques to supplement traditional one-by-one nose counting.

For one thing, they believe that many people (the poor, recent immigrants, both legal and illegal, and members of various minority groups) are extremely distrustful of the Census Bureau (along with all other government agencies). These people may fear that if they respond to the census, something bad is likely to happen to them. So, they don't respond to the mass mailings, they don't answer the door when the canvassers knock, and they end up not getting counted. Not every hummingbird is going to display its ruby throat to the enterprising bird-watcher, and not everyone wishes to be known to the Census Bureau. But, hummingbird or person, that doesn't mean they shouldn't be counted anyway. In addition to those people missed during the census (the undercount), there are those, including many college students, who are counted twice (the overcount).

The argument then goes further to say that an old-fashioned nose count (mail-in surveys and door-to-door canvassing) is going to miss (undercount) a much larger percentage of poor people, immigrants, and minorities, many of whom live in the inner cities and other low-income areas. The inhabitants of these areas will thus miss out on much-needed tax revenues and will be underrepresented in Congress when redistricting takes place. This could then become a vicious cycle. Feedback amplification? You bet!

Census directors have surely always been aware that the census fails to count everyone. In 1940, for example, 3% more draft-age men overall, and 13% more draft-age black men, showed up for the draft pool than had been tallied in the 1940 census. To fix these sorts of problems, the Census Bureau came up with a proposal, which it had intended to implement in the 2000 census, to replace the traditional nose count with a system wherein traditional census techniques are "enhanced" by statistical sampling. Is this true "measurement"? Not if you ask most Republicans in Congress. The whole thing turned into a classic case of elected officials, both Republican and Democrat, lining up on a measurement issue based not on the facts but on their political beliefs. The Democrats, by and large, were in favor of the Census Bureau's sampling plan. Of course, were this plan to be adopted, it would likely have resulted in increased inner-city population counts, and thus potentially more congressional districts for these areas that are normally heavily Democratic. Nonsense, most Republicans countered. A traditional nose count is mandated by the Constitution. In addition, Republicans argued that sampling techniques are technically suspect. They must have felt vindicated by Florida 2000. But then again, so must have the Democrats. That particular election didn't inspire much confidence in either statistical sampling or traditional nose counting.

As with global warming, and the age of the Earth and intelligence testing before it, this is yet another case where people's opinions about what should be a straightforward measurement issue are hopelessly intertwined with their political, religious, or cultural beliefs. This is so obvious as to be laughable—if it weren't so sad. But beyond that, this example also shows that the line between classic measurement and measurement derived through computer-based modeling is blurring. The goal of a census is simple: figure out how many people there are and where they live. If we can get a better answer using sampling and computer modeling, fine. If the best answer is still obtained the old-fashioned way, then at least for the time being, we'd better stick with that. It is unfortunate, but hardly surprising, that we as a society are unable to evaluate this situation dispassionately.

Computers are and will remain a crucial part of the measurement revolution, and an essential part of many vitally important measurement technologies, such as GPS, MRI, and digital photography. Computers don't just make measurement easier and more convenient. Without them, many of these technologies simply wouldn't be possible.

CHAPTER 10

HOW

FUNNY

IS

THAT?

Knowledge Without Measurement?

You can discover no measure, no weight, no form of calculation, to which you can refer your judgments in order to give them absolute certainty. In our art there exists no certainty except in our sensations.
—Hippocrates (ca. 460–ca. 377 BC)

A few honest men are better than numbers.
—Oliver Cromwell (letter, 1643)

The Measurement Cycle

The December 26, 2004, earthquake and tsunami in the Indian Ocean was one of the greatest natural disasters in recorded history. Over 150,000 people were killed and more than half a million were injured. In the aftermath of this calamity, there has been much discussion of tsunami warning systems. The history of our attempts at the measurement of tsunamis in order to provide such warnings serves as a good example of what I believe we can think of as the three stages in the "measurement cycle." From a broad consideration of all the various measurement techniques described throughout this book, it seems to me that this measurement cycle is more or less repeated each time measurement finds its way into a new area—such as the measurement of tsunamis for the purposes of developing warning systems.

This cycle consists of three stages of varying lengths of time: First there is the premeasurement era, which is characterized by ignorance, or at the very least by "meager and unsatisfactory knowledge." Decision-making during this stage is governed not by reasoned analysis but rather, at best, by vague hunches and the powers of persuasion, or, at worst, by nothing at all. The period of meager and unsatisfactory knowledge that characterizes the first stage of the measurement cycle is followed by the second stage: the development of measurement—through inspiration, innovative research, trial and error, and so on. Finally, there is the third stage: the application of measurement, and its further development and refinement.

Tsunamis are usually caused by underwater earthquakes, although they can have other causes, such as volcanoes, landslides, or meteor strikes. The earliest tsunami warning systems were developed in Hawaii in the 1920s. Prior to that, there was very little way to know when a tsunami might strike. Our ignorance was profound, but not complete. On earthquake-prone islands, such as in Japan, those who dwell near the shore have known for centuries to seek higher ground just after an earthquake has been felt. Experience taught that a tsunami might only be minutes away. However, tsunamis can strike coastlines even when the epicenter of an earthquake is so far away that its tremors are so small as to be imperceptible to humans. Following the Indian Ocean tsunami, some attention has been paid to the "sixth sense" that some animals seemed to have that warned them of this tsunami—it appears that relatively few land animals were killed by the tsunami. Do animals have internal seismographs more sensitive than we humans? We remain largely ignorant of whatever the animals may have been measuring internally that warned them of impending danger.

The early tsunami warning systems relied solely on the detection of earthquakes, and not on the tsunamis themselves. Seismographs measure the magnitude of earthquakes by quantifying the magnitude of the seismic waves that travel though the Earth in the wake of an earthquake (or a huge explosion like a volcanic eruption). When an earthquake whose epicenter is underwater is detected, warnings can be delivered quickly, because seismic waves travel through the ground 30 to 60 times faster than tsunami waves travel through water. Seismic wave speeds vary, but 8 kilometers per second (roughly 29,000 kilometers or 18,000 miles per hour) is a typical number. At that speed, a seismic wave would travel from New York to Los Angeles in about 10 minutes. Since tsunami waves only travel at 500–1,000 kilometers per hour (about 300–600 miles per hour) in open water, warning systems that rely on seismic waves generally allow enough time to warn affected populations. For example, a major earthquake off the coast of California could generate tsunamis traveling east that would hit California in minutes, but the waves traveling west would not arrive in Hawaii or Japan until hours later. The related, crucial issue of ensuring that people near the shore in affected areas actually receive the warnings in time is a separate question—the point here is that seismic warning systems, because of the difference in wave speeds, have the potential to allow such timely warnings. Seismic measurements allow quick calculation of an earthquake's epicenter, which then allows calculation of where, and when, any tsunamis that were created would hit land.

The earliest applications of seismic-based tsunami warning systems represents in this case the beginning of the second stage of the measurement cycle: development. The measurement of seismic waves was first demonstrated way back in 132 AD, when Zhang Heng of China invented the first seismograph. The development stage of tsunami measurement might thus be thought to date back to 132, although it appears that no one thought to apply seismic wave measurement to tsunamis until the 1920s.

One problem with tsunami warning systems based solely on seismic waves, however, is the likelihood of false alarms. Not all undersea earthquakes result in tsunamis. Up to two-thirds of all tsunami warnings based on seismic data alone are false alarms. With such a high rate of false alarms, such warnings risk being ignored, which is a huge practical problem. To lower the rate of false alarms, seismic data have been combined with oceanographic data to create a much more sophisticated system, in which the risk of false alarms is nearly zero. Adding the oceanographic component to the measurement of tsunamis represents a significant milestone in the development stage of the measurement cycle.

The oceanographic component of such a system utilizes undersea pressure sensors that detect increases in pressure when a tsunami wave passes above the sensor. The pressure sensor then sends a signal to a buoy on the ocean surface, which sends a signal to a satellite, which passes the warning on to a warning center, which notifies those in the path of the tsunami. Such a system exists in the Pacific Ocean—the Pacific Tsunami Warning Center. In the wake of the December 26, 2004, disaster, a similar system will surely be installed in the Indian Ocean. Other measurement techniques are being developed as well. Upon reviewing satellite radar data after the Indian Ocean tsunami, scientists from the National Oceanic and Atmospheric Administration determined that these data included quite accurate measurements of tsunami wave heights. In the open water, away from shore, tsunami waves are frequently less than one meter high, yet these radar satellites can directly measure their presence. The potential for satellite-based tsunami warning systems is under evaluation by the NOAA. The development and refinement of tsunami measurement and warning systems, both ocean-based and satellite-based, will continue. The stakes, as we now know, are high.

Meager and Unsatisfactory Knowledge— and the Power of Persuasion

Before measurement, that is, before a system for numerically characterizing something was developed, what do we have? In a word, ignorance. In the case of tsunamis, the situation could hardly be characterized as "ignorance is bliss" but perhaps more as "ignorance of ignorance," which has been characterized as the real foe of knowledge. Over the centuries, shore-dwellers learned to head for high ground after an earthquake, but this wasn't good enough. In many cases, there was simply no warning.

The same sort of thing applies in situations where measurement is needed to help us decide among alternatives. Such decisions might involve competing designs for a music hall, investment alternatives for an entrepreneur, devices for detecting bombs in luggage or freight, or the value of a professional athlete. Imagine a meeting between the legendary New York Yankees manager Joe Torre and his boss, the infamous Yankee owner George Steinbrenner. It is the off-season, and they are trying to decide how best to invest the boss's millions in the free agent market. Which earnest young slugger or fire-balling pitcher will get the nod? Without the ability to measure the performance of these prospective pinstripers, how could Joe and George possibly arrive at a decision? Bereft of the early contributions of Henry Chadwick (batting averages, earned run averages, and the like), to say nothing of the wondrously complex and insightful measurements of sabermetrics, Joe and George would be left only with their own considerable powers of persuasion. Having the appropriate measurements at hand does not in this case obviate the need for persuasion, there being as yet no perfect algorithm for selecting the best baseball player. But without the measurements, one is left only with anecdote, mythology, and who can yell the loudest. In that case, my money would be on Steinbrenner.

Development, Application, and Acceptance of Measurement

At some point, for some reason, someone decides that something can be measured that wasn't measured before. Zhang Heng was motivated by the frequent trembling of the ground in his native China. After several serious quakes, Zhang, who also invented an odometer, developed an instrument consisting of a large bronze jug with a pendulum inside it. During an earthquake, seismic waves move the pendulum, which pushes one of several levers connected to dragon's-mouth gates that encircle the jug. The lever opens one of the mouths, releasing a ball that drops to sound an alarm. The open dragon mouth points in the direction of the quake, notifying the emperor. Thus was born the science of seismography with its attendant measurement techniques.

Bill James and the other developers of sabermetrics in baseball must have said to themselves, "How long are we going to 'play hunches' in this business? There are so many ways to measure baseball players. It's high time someone figured out which of those ways are the best predictors of future performance." And so they did.

The development stage, with the requisite inspiration and innovation, is the hard part. The rest is relatively easy. In the free marketplace of measurement, new measures are always coming and going. If they are useful, they get used. If not, they are relegated to the metrological dustbin. Seismographs and sabermetrics have passed the test. The love meter, which I foisted upon you in Chapter 1, is unlikely to do so.

Baseball has not always, needless to say, had access to what we now call sabermetrics. What passed for the state of the art in performance measurement for baseball players used to be things like batting average and earned-run average—the kinds of measurements that are still, in the minds of the average fan, the most important. Baseball sabermetricians have moved beyond that, however. Different measurements are now used by those whose careers depend on selecting and developing the best talent. Better and better ways to measure baseball performance continue to be developed.

Knowledge Without Measurement

There is a certain kind of measurement that takes place in the mind, and not (yet) in any tangible way. We do not yet know how to quantify it or make it tangible. A craftsman creates something by hand. Someone, perhaps a prehistoric man, cuts up a stone to make a knife blade. How does he know where to make his cut? He looks at the stone, the visual information gets processed (measured inside the craftsman's head), and he decides where to cut the stone to get the best knife blade. He has perhaps made hundreds of good blades (along with thousands of bad ones), and his eye has become finely calibrated.

This sort of thing still goes on today. We do it all the time, dozens or perhaps hundreds of times per day. We measure things "inside our heads." We measure out prepared food portions, the ingredients for recipes, and dollops of toothpaste or shampoo. Lance Armstrong is said to weigh his food portions on an electronic scale before he consumes them. For most of the rest of us, the eyeball method is good enough. Engineers (particularly those of a certain age) often speak of a colleague who has "a calibrated eyeball"—who can look at a manufactured component and tell you if it is a hundredth of an inch too long, or if a drilled hole is off-center by a similar amount.

Both the stone cutter and the engineer with the calibrated eyeball are able to do what they do because they have practiced it. This really *is* measurement, but it isn't tangible, at least not yet. Someday, maybe we'll be able to attach a cable to a port that is wired into the base of our brains and "download" the virtual measurements we make in our heads every day. What would we find if we could do this?

Essentially, this is what we have succeeded in making computerized instruments do when we have them measure things. Using a laser "vision" system, a computer can measure whether a component is indeed too long, and, if so, by how much. Computerized part-inspection systems that can do this are quite common. You might say that the computer is simply imitating what a human being would do in this case, with several important differences: the computerized system is much faster, more accurate, more reliable, and its measurements can be recorded and saved.

What about the amazing ability of the human being to recognize faces? The average person can distinguish the facial appearances of thousands of individuals. Someone walks into the room, and you instantly recognize that person. You may not remember their name or anything else about them, but you know that you know them. This happens to me all the time with my former students. When I run into them, I remember their names, if I'm lucky, but even when I don't, I nearly always remember their faces.

What do we see, or more to the point, what do we measure, when we recognize someone's face like this—when we measure it in our heads? Computers, phenomenal measuring machines though they are, have a great deal of difficulty performing facial recognition with anywhere near the facility of the average person. If a computer can beat the greatest chess players in the world, why shouldn't it be able to reliably perform such a seemingly simple, mundane function as recognizing a face? Computers *are* getting much better at recognizing faces. They can now do a creditable job of scanning passengers' faces at airports, looking for known terrorism suspects, for example. It would seem that there is nothing fundamentally different between a computerized system for measuring the dimensions of parts in a manufacturing plant and one that measures and matches faces at the airport. It's all "measurement." When people do it, it's intangible, but we can make it tangible with computers and the right instrumentation. It is likely that we simply haven't hit upon exactly the right set of measurements when it comes to computerized recognition of faces in order to make these systems as fast and reliable as human "face recognition systems."

"Measuring in our heads" really is measurement. If we haven't figured out how to write it down, or to otherwise render it quantitatively, it is only a matter of time before we do.

Taking the Measure of the Old Masters

As an example, have you ever seen a painting whose intricate detail and flawless perspective reminded you of a photograph? The contemporary British artist David Hockney has. In his book *Secret Knowledge: Rediscover-*

ing the Lost Techniques of the Old Masters, Hockney proposes controversial theories about how such paintings were created. Through careful observation of the canvases of some of the old masters of painting (including Ingres, Caravaggio, and many others, mostly between the years 1300 and 1800), Hockney has come to believe that these paintings were created with the aid of optical devices such as convex mirrors, lenses, and a device known as the camera obscura. These optical devices would have allowed artists from that era to create photograph-like images on their canvasses—but they do not involve the use of photographic film, because it had not yet been invented. The images created by these optical devices are fleeting—unless someone were to trace over the image. Using these optical devices, a highly skilled artist might be able to capture a level of detail and a precise perspective otherwise practically unobtainable.

What did Hockney measure to derive this knowledge? In the beginning, he relied on his artist's eye to make the initial highly critical, but nonquantitative, observations that led to his theories. Certain features of various works of art have a certain look. Paintings that Hockney believes predated the use of optical devices look awkward, their perspectives are often flat, and details appear to be groped for. Later paintings and drawings, whose creators Hockney believes used optical aids, show supreme command of detail and perspective. (Hockney notices many other things as well. Why, for example, are so many figures in paintings from a certain era shown holding drinking glasses in their left hands? Hockney believes it is because a lens reversed the image on the canvas—many of these paintings happened to be created about the time of the invention of the lens.) Reading Hockney's book, I was convinced that he is onto something. His qualitative descriptions, juxtaposed with reproductions of the paintings, are compelling.

Hockney's qualitative descriptions do not represent knowledge of a "meager and unsatisfactory" kind. Nevertheless, the physicist Charles Falco, who read early accounts of Hockney's theories, offered to help the artist measure and quantify certain aspects of the paintings in ways that ratified Hockney's qualitative ideas.

An example of the measurements Falco was able to make concerns the use of "multiple perspectives" in some paintings. To aid in the creation of a painting of someone's face using mirrors and lenses, a single perspective would probably suffice. That is, the optical devices could be set up and left in one fixed position, and they would provide the artist with an adequately focused image of the entire face. To create a painting with a larger scope, such as a full-length portrait, or a painting with several figures in it, would require repositioning the optical devices several times. Thus, various parts of the painting would have slightly varying perspectives. Falco showed this by measuring lines of perspective on some paintings and showing how they did not match. These mismatched lines of perspective, when superimposed on reproductions of the paintings in Hockney's book, provide graphic (measured) evidence to support the use of optical devices in the creation of these paintings.

Hockney presumably entitled his book *Secret Knowledge* because the knowledge of how the old masters had used optical instruments to create such brilliant paintings had been their, the artists', secret, and because that secret had gone undiscovered for so long. But there is another way to look at the title *Secret Knowledge*. The knowledge that Hockney himself developed through countless hours of observation of these old paintings, and through his own experiments with mirrors and lenses, was in some sense "secret." Hockney knew it—he *knew* that these optical devices had been used to create these paintings. But how could he convince the rest of us? How to let us in on his own secret knowledge? By writing a brilliant book, for one thing, full of persuasive arguments and tons of illuminating illustrations. But the quantitative contributions of his physicist colleague should not be underestimated.

Hockney's writings, not surprisingly, have set in motion a vigorous debate in both the artistic and academic communities. Much (but not all) of this debate has been carried out using the customary rules, language, and tools of a scientific inquiry. When a scientific thesis is called into question, experiments and the measurements therein are analyzed and offered as evidence on one side or the other. Those on the other side of what is sometimes referred to as the Hockney/Falco thesis, such as David Stork of Stanford, interpret the laws of optics and the results of their own experiments (in some cases similar to those of Falco) to refute, painting by painting, the claims of Hockney and Falco. They use measurement and its interpretation to validate their claims, just as Falco does.

The reputation of its author (Hockney's art speaks for itself) undoubtedly has something to do with the popularity of *Secret Knowledge*, as well as with the liveliness of the ongoing debate. Stork admits that his first reaction to the Hockney/Falco thesis was, "this is pretty cool." His subsequent doubts were brought on by the results of his purely scientific inquiries into Hockney's work. The things that Hockney originally "measured in his head" are now being debated using more tangible, traditional, though no less creative, measurements.

There are other forms of knowledge without measurement, or without much of it, anyway. Let's press on to the acknowledged champion, the 500-pound gorilla of qualitative knowledge—the law.

The Law

Setting aside (albeit reluctantly) all the jokes about the legal profession, most of us would agree that the average lawyer is a "knowledgeable person." Among other things, she knows the provisions of the various laws that make up her specialty within the legal profession, such as bankruptcy law, criminal law, family law, or environmental law. Much of this is straightforwardly measurable. Every lawyer will tell you, however, that the percentage of cases covered by actual, existing laws (what a lawyer would call "black-letter law") is small—which is a measurement in itself, to be sure. The reason we have lawyers and judges, after all, is primarily to handle on our behalf the vast majority of cases for which existing statutes are inadequate.

Perhaps no legal issue is currently more controversial than same-sex marriage. In February 2004, a court ruling in Massachusetts mandated that state's legislature to change Massachusetts law to allow same-sex marriages in that state. For the first time, a man would be able to legally marry another man, and a woman another woman. This precipitated a series of events that eventually, according to some observers, ensured the reelection of President George W. Bush. In the United States, as in most countries, marriage has "traditionally" been defined as a legal union between one man and one woman. Traditionally, perhaps, but not by statute. The lack of precise laws defining who could participate in a legal marriage is what allowed this whole controversy to gain so much momentum. There are, I believe, compelling arguments on both sides of the same-sex marriage debate. Be that as it may, the number of states that have passed constitutional amendments or statutes defining marriage as between one man and one woman is on the rise. With these statutes and amendments in place, and once the dust has

settled on legal challenges to them, questions will nonetheless remain—
such as the legal definitions of "man" and "woman." Even a genetic definition is not unambiguous here. It is exceedingly rare, but possible, for someone with two X chromosomes (the normal configuration for a female) to
appear male, or for someone with one X and one Y chromosome (normally
male) to appear female. In addition, how will states handle individuals who
have legally undergone sex-change operations?

Defining marriage and gender and other questions involved in this debate ensure that a great many lawyers will stay busy for a long while. What
do lawyers do in cases like these? For one thing, they search for precedents.
These are previous cases that have been ruled on by judges or juries and
that bear some similarity to the case at hand. Knowledge of precedents is in
some sense measurable. Sophisticated searches of online databases provide
the lawyer with relevant cases, which must still be interpreted.

But these precedents are always at least a little bit different from the case
at hand. The entire body of facts in any case can never be precisely matched
by the facts in a preceding case. Once a lawyer has studied the available
precedents, she must convince a judge or jury why the rulings and decisions
made in these precedents should (or should not) apply to the matter under
litigation. How does the lawyer do this? By being persuasive, by making her
client appear sympathetic (particularly if a jury is involved), by creating
doubt, and so on. If it all sounds a bit like the first stage of the measurement
cycle (before measurement exists), it is!

No profession, it would seem at first, is less amenable to measurement
than the law. It is not just that the kinds of knowledge involved in the law
appear to be intrinsically resistant to measurement; our legal system itself
seems to have been specifically designed to be nonquantifiable.

I was once retained by an attorney as an expert witness in a lawsuit
involving an automobile accident. The attorney asked me to do several
things. First, he wanted me to evaluate a certain component on the vehicle
in question and describe that component and its operation to a jury in
layman's terms. No problem. Second, he wanted me to investigate several
alternative technologies for that component—products of other companies—and to give my opinions as to whether those technologies could have
been implemented on the car in question, and if so at what cost. No
problem there, either. Finally, he wanted me state whether the car, at the

time of the accident, had been in an "unreasonably dangerous condition." *That* presented a problem! I informed the attorney I felt uncomfortable, and indeed unwilling, to make such an evaluation, since I have no idea how to measure whether something is "unreasonably dangerous." The attorney responded calmly and with lawyerly assurance that it was really quite simple. With a great many flourishes and a lot of talk of the kind lawyers are famous for, he instructed me as to how I might evaluate the unreasonableness of the danger inherent in the car in question at the time of its accident—but he never did convince me. Engineers generally just don't feel comfortable with this sort of thing.

Let's say I propose to design a new bridge for the highway department. I show up at their offices to pitch my plans, briefcase and laptop in hand. How should I expect them to respond if I tell them that my bridge will be "reasonably strong," that it will be able to carry a "reasonable traffic load" in "reasonable safety," that it will last a "reasonably long time," and, finally, that it will be "reasonably inexpensive." With this sort of approach, I am reasonably sure that I would not get the job.

There are certain aspects of the law that are *intentionally* nonquantifiable. For example, there is the famous phrase "beyond a reasonable doubt" (there's that "reasonable" word again). In most criminal jury trials, the jury must determine if the defendant is guilty beyond a reasonable doubt. But judges are actually forbidden to interpret for a jury what is meant by a "reasonable doubt." Our legal system is founded on an intentionally unquantifiable principle. Should you ever undergo a criminal trial by a jury of your peers, your fate would depend on how 12 different persons interpret the phrase "reasonable doubt" with respect to your guilt. Their decision to convict must in most cases be unanimous, thus presumably making up for the possibility that a few jurors might hold unreasonable views on what constitutes reasonable doubt.

But, in many ways, and more and more each day, the law and our legal system are being quantified. More and more lawsuits, criminal trials, and other legal proceedings these days seem to hinge on quantifiable scientific evidence of one kind or another.

The polygraph, or lie detector test, is a famous and mostly failed attempt to measure the truthfulness of a witness. However, there are all kinds of good examples where quantitative methods have been successfully integrated into our legal system. For example, jury selection for criminal and civil trials is becoming more and more quantitative all the time. In a manner that mirrors the way professional athletes are selected, a lawyer's "gut feel" for which prospective jurors should be eliminated or retained is slowly being replaced by quantitative evaluations provided by sophisticated jury consultants. To be sure, such services are beyond the reach of all but the wealthiest plaintiffs and defendants (such as large corporations). But the way jurors are selected is beginning to resemble the way in which athletes are selected in a professional sports draft—by the numbers.

Scientific testimony creates a number of difficulties for our legal system. Among them is the question of which scientific experts, and which scientific theories, are even admissible as evidence in the first place. In the case of *Daubert v. Merrell Dow Pharmaceuticals, Inc.*, the U.S. Supreme Court ruled in 1993 for standards requiring minimum levels of relevance and reliability for the use of scientific or technical expert testimony in the courts, saying that testimony that did not meet such standards would not be allowed.

Subsequent developments have resulted in what are commonly referred to now as the *Daubert* standards. Among these standards are the following questions that judges should consider in determining whether an area of science is reliable enough to be allowed as evidence in court:

- Is the evidence based on a testable theory or technique?
- Has the theory or technique been peer reviewed?
- Does the technique have a known error rate and standards controlling its operation?
- Is the underlying science generally accepted?

Let's say a chemical company is being sued because some of its workers got cancer. The plaintiffs are likely to hire experts who will testify that exposure to certain levels of certain chemicals in the defendant's factory increases the risks of cancer, based on certain scientific theories and experimental measurements. *Daubert* says the theories and techniques used to create this evidence must be "based on a testable theory or technique" and "have a known error rate and standards controlling their operation." These are measurable things.

The *Daubert* standards are somewhat controversial in legal circles and have generated a vigorous ongoing debate, as well as numerous court challenges (creating more work for lawyers everywhere, it would seem). For example, fingerprint evidence generally does not meet the *Daubert* standards, because of problems with "known error rate." DNA evidence generally does meet the standards. Why are fingerprints still widely accepted as evidence, then? Well . . .

Putting Your Finger on It

Once upon a time, fingerprinting technology revolutionized crime fighting and criminal law. The uniqueness of fingerprints had been known or at least suspected since ancient times. In ancient China, fingerprints were used on official documents, and there is some evidence that criminals had their fingerprints recorded in both China and Egypt. In other ancient societies, fingerprints were used to seal business deals.

Fingerprints were first used as legal evidence in a trial in 1892, in Argentina. Their first use in the United States was not until 1911. A Chicago man, Thomas Jennings, had been convicted of murder based largely on fingerprint evidence left in wet paint on a window railing where the murderer evidently broke into and entered the home of the victim. Upon appeal to the Illinois Supreme Court, the conviction was upheld. "When photography was first introduced, it was seriously questioned whether pictures thus created could properly be introduced in evidence, but this method of proof, as well as by means of X-rays and the microscope, is now admitted without question," the court wrote. "[W]hy does not this record justify the admission of this fingerprint testimony under common law rules of evidence?" Jennings was executed for this murder in February 1912.

Before 1896, several important uses for fingerprints had been developed. It was already known that fingerprints did not change over a person's lifetime, and it was known or at least strongly suspected that an individual's fingerprints were unique. Fingerprints had been used for identification, for example, by illiterate persons who could not sign their names. Several crimes had even been solved through the use of fingerprints—but only when there was a relatively small group of suspects. To make fingerprints really useful, a system was necessary whereby fingerprints could be rapidly compared to a large database of prints. This was well before the dawn of the computer era.

Enter the Englishman Edward R. Henry, who developed the system, in 1896, that is still in use throughout most of the world today. Henry built on the work of the highly versatile Francis Galton and others. Various geometric aspects of fingerprints are still sometimes referred to as "Galton's details." But it was Henry who struck upon a practical classification scheme. Henry first broke fingerprints down into their three basic categories: loops, arches, and whorls (along with a none-of-above-category called "accidentals"). About 65% of all prints are loops, or patterns of ridges that begin and end on the same side of a finger. Arches (5%) are patterns of ridges that begin and end on opposite sides of the finger, and whorls (30%) look like a bull's eye. Eight of my own ten fingerprints are loops, and the other two are whorls, so I'm not far from the average.

Within each of the three categories, Henry identified subcategories. There are radial and ulnar loops, for example, depending on whether they are aimed toward the thumb or the little finger. After that, the Henry system counts the ridges in between various features in a print. When a given person's fingerprints have been thus categorized, a "Henry code" will have been generated, which is a combination of letters (indicating the type of prints on various fingers) and numbers (indicating the number of ridges in various places). Thus, the Henry system gives a unique code to each person, not to each individual finger's print. The system is ideal for use when a complete system of prints is available, such as for criminals in custody, applicants for government or other jobs requiring security clearance, and so on. For identifying who made a single print, say a thumb print, at a crime scene, the Henry system is less useful. A separate system for single prints was invented in 1930 at Scotland Yard.

With the Henry system and, later, the Scotland Yard system for individual prints, the use of fingerprints in crime fighting and elsewhere grew rapidly. J. Edgar Hoover built his early career heading up the "Identification Division" of the FBI, which he would later direct for so many years. Fingerprint technology today is a fascinating mélange of computerized quantification and old-fashioned art. (This is why it is still controversial in some legal circles, as we shall see.)

The first attempts to computerize fingerprint identification began in 1967. By 1979, police and other law enforcement officials could query a computerized database by name or by various fingerprint characteristics. Today's versions of such databases can search more than a thousand prints a second.

When a woman was murdered in San Francisco in 1978, police combed through over 300,000 fingerprint records, manually and over countless hours, without finding a match to several prints left at the crime scene. In 1985, the computerized system found the murderer in six minutes. But it wasn't quite that simple. First, the prints from the crime scene were scanned electronically and compared with the digital archive. The computer created a relatively short list of possible matches—say a dozen or so, and then a police officer expert in fingerprint identification eyeballed those prints to quickly nail down the murderer.

And therein lies the problem. There is still no uniform standard for making that final determination. Some print examiners use a system that requires a certain number of "points of similarity," while others rely only on their "overall impression" in making the final identification. The problem is worse when only a partial print is available or when it is smudged.

Since there is no uniform standard, there is at least potentially a problem with the *Daubert* standards mentioned earlier. A decision based on an expert's "overall impression" of a fingerprint seems unlikely to be "based on a testable theory or technique," as required by the *Daubert* standards. Its error rate will likely be unknown as well. The "underlying science," however, is clearly generally accepted and has been for a long time. It thus appears that fingerprints have been grandfathered in as a generally admissible form of quantitative scientific evidence, even though they appear not to satisfy many aspects of the *Daubert* standards.

Mental Health

Medicine is a field where measurement has made tremendous contributions. We've come a long way since the day when Hippocrates could tell his fellow physicians (with a straight face, presumably) that "you can discover no measure, no weight, no form of calculation, to which you can refer your judgments in order to give them absolute certainty."

Mental health care has also made great advances. Many conditions are now acknowledged to be mental illnesses that were once thought to be simply "attitude problems." These conditions, once properly diagnosed, can be effectively treated in a variety of ways, including drug therapy and counseling. Illnesses of the mind, however, are generally not given the same consideration and care that more conventional diseases of the body receive. As one result, in the United States, health insurance for mental health care is likely to be much less comprehensive than coverage for other, more conventional ailments.

There are probably many reasons why mental health care lags behind more traditional medicine. For one thing, the old-fashioned perception that mental health disorders are somehow the fault of the person suffering from them is still widespread. Most of us have nothing but sympathy for a cancer patient, but someone who has been diagnosed with bipolar disorder seems to many of us to fit in a different category. We may have sympathy for such a person, but in the back of our minds there may be some doubt as to whether this is a "real" disease or condition. We may also feel that this person could probably be "normal" if only he or she had more willpower or self-discipline.

Diagnosing Mental Illness and Conventional Illness

Another important reason, I believe, for the difference in the perception of, the care of, and the research expenditures for mental illness relative to conventional illness relates to measurement. Conventional (nonmental) diseases can generally be measured precisely these days, giving rise to diagnoses that are typically objective. By contrast, the diagnosis of mental illness seems not to have advanced much beyond the state of the art of Hippocrates's day: "In our art there exists no certainty except in our sensations."

Diabetes, for example, is a disorder characterized by high levels of blood sugar. This disease is quite common and serious; diabetes can result in long-term complications that can affect the kidneys, eyes, heart, blood vessels, and nervous system. It contributes to the premature deaths of a great many people. While the causes of diabetes are often mysterious, its diagnosis is relatively straightforward. If a doctor suspects you have diabetes, there are several tests that can be used to measure whether you do in fact have any of the several forms of this disease. The most widely used test involves measuring a patient's blood sugar level after the patient has fasted for 8–12 hours. A typical range for fasting blood sugar levels is 70–110 mg/dL. If a value above 140 mg/dL is observed on at least two separate tests, the patient probably has diabetes.

The second common measurement for diabetes is the oral glucose tolerance test. The patient, having fasted for 10–16 hours, is given an initial blood sugar test and then drinks a glucose solution containing 75 grams of glucose, or 100 grams for pregnant women. The patient's blood sugar is then measured 30 minutes, 1 hour, 2 hours, and 3 hours after drinking the glucose solution. In a nondiabetic patient, blood glucose levels will at first be quite high, but they will return to pre-test levels relatively quickly. If the patient is diabetic, blood sugar levels will remain elevated for much longer. There are quantitative standards for how high and for how long blood sugar levels must persist in order to warrant a diagnosis of diabetes.

Threshold blood sugar levels for a diagnosis of diabetes can be changed as researchers learn more about the disease and continue to study statistical data from both diabetics and nondiabetics, and there will always be difficult-to-diagnose borderline cases, but it remains true that diabetes, in its several forms, is a highly measurable disease. Blood sugar level is so routinely measurable that many diabetics and other patients do their own measurements at home. It is almost as routine as taking your temperature.

Contrast the diagnosis of diabetes with that of a mental condition known as "borderline personality disorder." This condition, believed to be relatively common, has been described as "a pervasive pattern of instability of interpersonal relationships, self-image, and affects, and marked impulsivity beginning by early adulthood and present in a variety of contexts."

According to the revised fourth edition of the American Psychiatric Association's *Diagnostic and Statistical Manual of Mental Disorders* (DSM-IV-TR), a diagnosis of borderline personality disorder is warranted if a patient exhibits at least five of nine "behavioral patterns," of which the following three are included: "frantic efforts to avoid real or imagined abandonment," "identity disturbance: markedly and persistently unstable self-image or sense of self," and "chronic feelings of emptiness."

The example of borderline personality disorder was chosen, as were three of the nine diagnosis criteria, to provide as graphic a contrast as possible between the seemingly vague, obviously qualitative criteria used to diagnose borderline personality disorder with the quantitative objectivity of a diagnosis of diabetes. But one really doesn't have to try that hard to find such a graphic comparison. Mental health, including the diagnosis and treatment of mental illnesses, remains a largely unquantifiable, subjective enterprise. Is our current knowledge of mental illness meager and unsatisfactory? Yes and no. On the one hand, the mental health industry is full of dedicated professionals who make a positive difference in the lives of their patients every day. But where, on the other hand, are the measurement techniques that will allow us to quantify the diagnosis of mental illness?

The first edition of the DSM (DSM-I) was published in 1952 in an attempt to bring some order to the chaos involved in diagnosing mental illness. DSM is widely accepted by psychiatrists and psychologists alike, but it is certainly not without controversy. There are those who claim that in their zeal to put a name on a patient's condition (a diagnosis) in accordance with DSM, clinicians can lose sight of what really matters—treating the patient and improving his or her life.

DSM is a classification system for mental illness. As noted earlier in this book, classification frequently precedes quantification. Although it will probably be a long time coming, the quantification of mental illnesses may indeed be on the way. Brain-imaging studies using functional MRI and similar techniques are shedding all kinds of new light on a wide variety of subjects, such as how we learn, how our brains develop, and mental illness, as briefly noted in the previous chapter. Clear differences, as revealed by these measurement techniques, exist between the brains of those who suffer from extreme cases of, for example, schizophrenia, and those with no signs of this disease. Development of these techniques as a diagnostic tool is in its infancy. It is difficult to predict where all this will lead, but if the measurement cycle introduced in this chapter is any guide, we shall soon begin to see the development and mainstream application of quantitative measurement techniques for diagnosing and treating at least some mental illnesses.

There is at least one measurement issue that bridges the gap between conventional medicine and mental health: how can we measure death? As the Terri Schiavo case demonstrated, modern medical technology can keep a heart beating well after its possessor has surrendered all or most of the other tangible signs we associate with human life; the ability to communicate, to feed oneself, to control bodily functions. Is such a person dead?

Death was historically defined by a lack of heartbeat and respiration. That has changed. Cardiopulmonary resuscitation (CPR), heart defibrillation, and other technologies routinely rescue persons whose hearts have stopped and who are no longer breathing. On the other hand, it is also possible for someone with a beating heart and functioning lungs to be considered dead. Today, a lack of heartbeat and respiration is sometimes referred to as "clinical death," which is sometimes reversible. This is to distinguish it from a cessation of all electrical activity in the brain, which is known as "brain death" and is considered irreversible. There are those who maintain that brain death should be defined more precisely as a lack of electrical activity in the neocortex region of the brain, and not the entire brain. But in the United States, the Uniform Definition of Death Act opts for the more conservative definition—death is the cessation of electrical activity in the entire brain.

Electrical activity in anything (from a thunderstorm to a computer to a brain) can be measured, and it turns out that measuring it in the human brain in order to determine death is not without its problems. Electroencephalograms (EEGs) can detect spurious signals in a brain when no activity is present, and it is also possible for a living brain to produce signals too weak to be detected by EEG. Some hospitals have adopted elaborate EEG protocols, including repeated testing in order to overcome these difficulties. But these are relatively straightforward measurement issues. They do not even begin to address the difficulties associated with other end-of-life questions, such as how to define a "persistent vegetative state" and under what conditions it is acceptable to remove someone from life support systems. Can the end of life itself be reduced to a set of simple measurements?

CHAPTER 11

FAITH, HOPE, AND LOVE

The Future of Measurement—and of Knowledge

And now faith, hope, and love abide, these three;
and the greatest of these is love.
—1 Corinthians 13:13

There is something about measurement, at some level, that we don't trust. Or maybe it isn't so much that we don't trust measurement. Perhaps we simply yearn for things in our lives that can't be measured, now or ever. Spiritual gifts such as faith, hope, and love certainly qualify. But when I read Saint Paul's letter to the Corinthians, I can't help but wonder how Paul could be so sure that "the greatest of these is love." Just what is he measuring here? It's not as though he wrote, "And now seven, twelve, and forty-five abide, these three; and the greatest of these is forty-five." I'm sorry, but I just can't help it. This is how I think about things! But even I am not immune to a sort of vague inner desire for things that can't be measured. I'm just not sure what those things are anymore.

Measuring Happiness

Prayer and its effects are measured all the time these days (and have been, dating back at least to Galton's treatise on the efficacy of prayer). Most modern studies investigate the link between prayer and healing—often finding a significant correlation. Cross prayer off the list of things that can't be measured.

Along with happiness. True happiness in life represents the ultimate goal for many of us. Yet a more slippery, difficult to define quality one could scarcely hope to find. You are either happy or you aren't, aren't you? How could you hope to measure that? Well, you can measure anything, as the psychologist Carol Rothwell and the self-styled "life coach" Pete Cohen have demonstrated. In an unpublished study commissioned by a vacation company (and reported by CNN and others), these researchers have concluded that happiness can be measured through a combination of personal characteristics P (outlook on life, adaptability, and resilience), existence E (health, friendships, and financial stability), and higher order characteristics H (self-esteem, expectations, and ambitions). Happiness, they found, is equal to $P + 5E + 3H$ and may be measured through a written survey.

Why measure happiness? To make money, for one thing. The sponsor of this study was almost certainly looking for ways in which to improve its business by understanding and quantifying how to make its customers happier. If you can measure happiness, you can study the effects of any number of factors on happiness. If your measurement is anywhere close to the thing itself (if it is close to being a homomorphism), then you have something of value.

Are Rothwell and Cohen onto something? Have they discovered a true law of nature—that which truly makes us happy? I doubt it. Does it really matter? Not really. Ask instead whether they have provided a useful service to their customer. At the very least, they have attempted to shine a light on the darkness perceived by their sponsor, the vacation company. They may not be measuring exactly the right thing, but they are measuring something.

Rothwell and Cohen's work might even be valuable for someone besides their client. Take me, for example. If I am feeling vaguely unhappy and am really not sure why, I might look at how they have classified and then measured happiness. Might my unhappiness result from a poor outlook on life? Or perhaps a friend has moved away? Or maybe I have given up on a long-term career goal and thus lowered my ambitions? I don't need to make too much of the actual numbers in their overall measurement for happiness—I can simply self-evaluate the individual aspects of "happiness" that go into their overall ranking. I needn't worry about how my happiness measures on a scale of zero to one hundred or anything like that. That overall quantitative measurement might make a lot of sense for Rothwell and Cohen's client, but personally, I'll give it a miss.

Awash in a Sea of Numbers—What to Do?

So here we are, awash in a sea of numbers. One by one, all the great mysteries of life seem to be yielding to the precise application of one sort of measuring stick or another. Where will it all end? Shall I arrive someday at the Pearly Gates, only to hear Saint Peter say, "I'm terribly sorry, the Goodness Index of your life was only a 69.4. We require at least a 70 to enter the Kingdom of Heaven—no exceptions. I'm afraid it's the Other Place for you." Will I be able to do an extra-credit assignment to raise my grade?

Measurement is no different from any of the other trappings of modern society. We can be its master or we can be its slave, but we can't make it go away. What kinds of things can we do to master our measurement culture? Most of my prescriptions for dealing with measurement are educational (I'm a teacher, after all). Thus, we have to learn about and understand the measurements that rule our lives. This means knowing where they come from (what's behind them—what's going on "under the hood"), knowing what they mean, what they don't mean, how they lead us, and how they can mislead us. It also means knowing when we don't know something—since nobody is a master of all forms of measurement.

Numeracy and the Cultivation of your Quantitative Side

Measurement is such a pervasive part of modern life that you might think it would tend to make us more rather than less numerate. But somehow we are insensitive or have been rendered numb by the explosion of quantitative information around us. You don't have to look too hard to find examples.

A college student of my acquaintance (of above-average intelligence, I hasten to add) recently asked me if it "made sense" that the fuel tank in his car, a small SUV, might hold 79 gallons of fuel. No, it didn't, I offered. (I'm still not entirely sure what precipitated this unusual question.) How on Earth, he wondered, was I able to answer his question so quickly? I asked him how much it cost him to fill the tank when it was almost empty. He replied that he had once paid nearly $45. At the current price of roughly $3 a gallon, I told him that meant his gas tank would hold about 15 gallons— which meant it would take more than five of his car's gas tanks to contain 79 gallons. (I also pointed out something he had somehow reached the age of 20 years without divining: that the flow meter on the side of the gasoline pump is calibrated in gallons as well as dollars.) If a good student like this (he may be your doctor someday) can be that innumerate, it is not alarmist to say that we have a long way to go preparing our children to be good citizens of our measurement culture.

It doesn't help that in large segments of American society, quantitative skills are looked upon with such disdain. Students who are good at math tend to keep that fact well hidden, lest they be labeled geeks or nerds. It is almost a point of pride for American students to "boast" that they aren't good at math. I even hear engineering students say this. Much as I would like to, I can't change that aspect of our culture. But it needs to change.

Beware Those Weighted Averages!

Part of being numerate is knowing when the numbers make sense. It doesn't make sense that a car's gas tank would hold 79 gallons, just as it doesn't make sense that an Internet investment scheme could double my money in only three weeks. And I'm not a big fan of weighted averages. They frequently don't make sense either.

My doctoral research was a study of the ways in which engineers make the "early decisions" when they are in the process of designing something. For example, let's say you want to design a new lawnmower. One of the first things you are going to have to decide is which basic type of lawnmower you are going to design. Among the early decisions you will need to make is whether the lawnmower will be a riding mower or a walk-behind mower (or perhaps even an autonomous unguided robotic mower), whether it will be gasoline, electric, or human-powered, and whether it will have reel-type or rotary blades. You may also be considering a cutting-edge (no pun intended) innovative new concept for grass-cutting, such as one that uses a laser beam to trim the grass or burns yard waste as its fuel. Since you only have the time and money to design and build one lawnmower, selecting the "concept" for your mower from among these possibilities is among the fundamental "early decisions" you will have to make. As part of my research, I evaluated various formal and informal techniques that engineers use to make such decisions.

One such process for "concept selection" involves systematically comparing the potential concepts (rotary blades vs. reel vs. laser . . .) in terms of the main things you want your new mower to do. For example, you want your mower to be inexpensive, lightweight, low-maintenance, and so on. You can facilitate the concept selection process by making a table and filling in your concepts along the top row and the things you want (low cost, etc.) in the left-hand column. You can then fill in each box in the rest of the table with your evaluation of how well or poorly a particular concept provides a particular want. There are various ways to do this, but I'll skip the details here (if you're a glutton for punishment, you could always read my doctoral dissertation).

What I like about the above technique (first developed by the late British aeronautical engineer Stuart Pugh) is that it helps reveal serious flaws in particular concepts very early in the design process, thus saving money and time. For example, the laser lawnmower is likely to have cost and safety disadvantages relative to more conventional concepts. What does all this have to do with measurement and weighted averages? Well, among the variations on the above matrix-method of concept selection are several that involve quantification of the various concepts through the use of weighting factors. Points are assigned to each concept in terms of each of the wants. (Once again there are several ways to do this—the details are not important.) Through observation of design teams who have used these methods, I have found that engineers often prefer to "quantify" the concepts with weighting factors (rather than simply using the matrix technique as a starting point for a qualitative debate about which concept to choose). I have even watched design teams carefully adjust the weighting factors in their concept matrix in order to ensure that their favorite concept ended up with the best score!

Weighted ranking systems like this, whether they involve lawnmower concepts, football teams, or universities, are almost infinitely adjustable. I could make a butter knife look like the best lawnmower concept by drastically increasing the weighting on "low cost" while decreasing it on "ease of use."

But weighted averages are here to stay, I'm afraid. It pains me not a little to confess that I use them myself in my professional life all the time. A grade of "A" in one of my courses might, for example, require a score of at least 90% based on a weighted average of test scores and homework and report grades. Mea culpa. Not to mention caveat emptor.

Cooking the Numbers

Sometimes, in our society, we use measurement to intimidate. Sometimes it's the other way around, and we use a lack of measurement to intimidate. As one example of the latter, consider cooking.

To many people, cooking is a mysterious art form. Some of this stems from the intentionally qualitative approach in many cookbooks; "season to taste," or, "stir until sauce thickens," and so on. You can't season to taste unless you already know what it's going to taste like. Thus, the novice chef is left out in the cold by such an approach. But there's no good reason for this, because cooking is merely chemistry. Sodium chloride (table salt) is an important ingredient in many chemical products, just as it is an important part of many food recipes. Can you imagine the recipe for making, for example, polyvinylchloride (PVC plastic) calling for "a pinch of sodium chloride"?

Attempts to demystify cooking include the Fannie Farmer approach. Fannie Farmer published her first cookbook in 1896 (today the Fannie Farmer tradition is carried on by down-to-earth cook and author Marion Cunningham). Fannie Farmer's 1896 book introduced the concept of "level measurement" in which everything is quantified. "Add enough flour to thicken the sauce" becomes instead, "Add ½ cup of flour." This quantitative approach is useful for the novice or uncertain chef. Following recipes like this, he will not be uncertain for long. Today, Marion Cunningham combines the quantitative and qualitative sides of cooking in her work. She continues to demystify cooking through the use of level measurements, while at the same time exhorting her readers to taste constantly and experiment with their recipes, and thus hone their qualitative, sensual cooking skills. Since the product of cooking goes in our mouths, we'll never be able to divorce the quantitative from the qualitative aspects of cooking. Taste is like that.

Taste is the most notoriously difficult of the senses to measure. Food factories, where products as diverse as ice cream, beer, and burritos are prepared, are highly engineered places that are often marvels of modern technology. I've toured many food factories, and I'm always impressed at the level of technology they employ (plus I love the free samples at the end of the tour). In the modern food plant, everything is antiseptically clean, ingredients are carefully (and automatically) measured, and temperatures and pressures are electronically controlled. The color, texture, and consistency of the product can be measured and thus controlled. But evaluating the product's taste stubbornly refuses to yield to high-tech instrumentation. For that, the old-fashioned taste tester is still a necessity. Taste testers in food plants are carefully chosen and often undergo months of training, only to be rejected when it becomes clear that their palates are not discriminating enough. A master taste tester ranks among the most valuable employees in any food plant and often has the power to reject an entire batch of product.

Will taste testing someday yield to the onslaught of measurement technology? Perhaps, but in the meantime, and even after that has happened, the taste tester has some lessons to teach all of us about our senses. Even as she provides us her quantitative, level-measurement recipes, Marion Cunningham exhorts us to cultivate the taste tester within. But why stop with taste?

Train Your Senses

These days, it is easy to make fun of Hippocrates when he says, "In our art, there exists no certainty except in our sensations." Doctors are now trained to trust their instruments—and the measurements those instruments provide—because their own sensations are in many cases inferior. To be sure, doctors are still trained to use their senses—a good physician can palpate a patient's body (i.e., examine it by touch) and discover things that the rest of us would never notice, and doctors still listen to our hearts and lungs through stethoscopes. But more and more of a physician's training is devoted to understanding and applying the results of the various measurements (x-ray, ultrasound, blood chemistry, MRI, and so on) that dominate modern medicine. These tools will clearly measure things that cannot be detected "in our sensations," and it is only natural that doctors devote more of their training to getting maximum utility out of these tools, and that their training of their own sensations thus suffers.

On the other hand, the senses we were born with, if properly developed, remain exquisitely powerful tools. An artist's eye is a wonderfully trained instrument. You can't draw, paint, or sculpt something if you can't see it— and I mean *really* see it. Artists are trained to see colors, shadows, perspectives, and details that simply don't exist for the rest of us. Betty Edwards makes this point in her excellent book *Drawing on the Right Side of the Brain.*

If you want to sharpen your sense of taste, learn to cook. If you want to sharpen your visual acuity, learn to draw. The engineers I know who draw the best are those who "see" the best. They see details in things the rest of us miss, and they can also visualize, for example, how a complicated three-dimensional machine (like an automatic transmission or a photocopier) is going to fit together before it is built. Computers make this task easier today, but an engineer's ability to visualize is still extremely important.

But as computers and measurement instrumentation become more and more powerful, easy to use, and (let's not forget) cheaper, it gets harder and harder for me as an educator to convince my students to develop their own senses—their internal measurement systems. (I suspect those who teach medical school must have the same problem.) Nonetheless, I still love teaching in the laboratory. In the lab, so many more of our senses come into play than, for example, in the lecture hall. We see, hear, smell, feel, and even taste in the lab. I love to help my students develop their abilities to hone their senses and powers of observation in the lab, but it's getting harder all the time. It seems to me that my students approach their laboratory studies much more passively than they did just a few years ago.

The type of learning that goes on in the traditional classroom lecture hall is generally passive; students listen, make notes, and hope to figure out what will be on the test. In the lecture hall, the student seems to say, passively, "Tell me what to write down." In the laboratory, these days, she seems to say, "Tell me where the electronic file with all the data on it will be." An experiment is performed, a set of powerful instruments attached to a computer gobbles up all the data, and an electronic file is created. The teaching laboratory has become little more than a device for creating that data file, armed with which the student will later (probably about 3 a.m.) crunch some numbers and write a report. That they might use the laboratory as an opportunity to sharpen their visual, aural, tactile, or other powers of observation seems to be lost on many if not most students (and more than a few of their professors, I'm afraid) nowadays.

This is something that I usually try to point out to students in the lab; that is, that the lab is a good place to learn how to observe. Having delivered one such homily, I remember reading the lab notebook of one of the better students in the class, where she had carefully noted down: "The eye is a powerful instrument if trained." I thought it odd that she chose to make a note of that. Had this never occurred to her before? Or did she maybe think that it might be on the test? "Which of the following is a powerful instrument if trained? (a) the eye, (b) the navel, (c) . . ." Perhaps she thought it was a profound observation that she should always remember.

We seem to have forgotten why it is important that our children study art or music. We should learn to draw or paint in elementary and high school, not because we want to or are likely to become artists (although a tiny fraction will), but because it helps us to see better. It helps us to develop one of the most powerful gifts with which most of us were born: our eyes. Likewise, learning music is to the ear what learning to draw is to the eye.

Getting in Touch with Your Inner Qualitative Child

I once read the results of a survey in which practicing engineers were asked which academic subjects they would prefer to study if they were somehow given the chance to relive their college years. Young engineers, those under 30, responded overwhelmingly that they would study more technical subjects—hard-core nuts-and-bolts, analytical, math-intensive engineering. Young engineers generally do the heavy lifting in most companies when it comes to intense analytical work, and their response to the survey thus makes sense; they are looking for more and better tools to help them solve the technical problems that dominate their professional lives. Those in the survey aged 30 to 40 most often responded that they would prefer to study a variety of business-related topics—things like management, finance, and marketing. Once again, this is not surprising, given that many engineers have become managers by that stage in their careers. Those over 40, however, were most likely to opt for serious study of the humanities—literature, philosophy, history, and the like. We engineers are perhaps a little slow. It would appear that it takes us until our 40s to realize that what really matters in life are people.

The late Ronald Reagan was adored and idolized by millions of Americans. I believe one reason for Reagan's remarkable popularity was that he appeared not to be profoundly influenced by the frenzy of measurement activities that are so deeply entrenched in modern politics—things like the polls and focus groups without which many politicians would be paralyzed. There are probably a variety of reasons why Reagan was able to lead the American people without a heavy reliance on these tools. For one thing, times have changed. Reagan left office in 1988, and the World Wide Web, for just one example, was not invented until 1990. For another thing, it's not as if Reagan's staff were ignorant of the quantitative tools of the political trade. The numbers were there for Reagan—he clearly knew (or at least his closest advisors knew) where the public stood on the issues of his day. He simply chose not to let the numbers dominate his policy decisions. Or at least that is the appearance he gave. We often hear politicians today say that they are guided by their principles, and not by the polls, but how many of us really believe them? We believed Reagan, however.

Rather than being influenced by polls, focus groups, and the like, Reagan appeared to lead with his heart, and through adherence to a few simple, qualitative principles. This, I believe, is prominent among the reasons why we loved and admired him so much. Even those who disagreed sharply with Reagan's policies (people like Tip O'Neill and Bill Clinton) admired the man for his adherence to his principles. Since Ronald Reagan, Americans seem to have become relentlessly more cynical about politics. We tend to believe that our leaders refuse to take a stand on anything until they have "run the numbers." We admire a politician with the guts not to live by the numbers. (But even though we admire him, we don't always elect him!)

Ignoring the Numbers

A colleague of mine is a professor at one of the top research universities in the United States. She recounted for me a discussion she once had with her department chairman, a brilliant researcher with an international reputation. The chairman stated flatly that he would under no circumstances bring on a graduate student to work under his tutelage unless that student had scored an 800 (the maximum score) on the quantitative (math) portion of the Graduate Record Exam (GRE). The chairman had the luxury of requiring such a high standard. His department being one of the top five in the nation in most rating schemes, there was never a shortage of student applicants boasting an 800 on the GRE math.

At this point my colleague (who had just received tenure and is in her own right a brilliant academic) asked her chairman if the same criterion should apply to prospective faculty members in the department. The chairman allowed as how it should—the standard should be at least as high for professors as for graduate students, shouldn't it? Then (and you could see this coming) my colleague gently informed her chairman that she had somehow slipped through the door having only achieved a 700 on the GRE math.

The moral of this story is simple. Quantitative criteria such as the GRE are important, but it is foolish to use them blindly and arbitrarily. We do it all the time, though. Making decisions like this (student admissions in this case, but there are other examples) based solely on quantitative criteria is tantalizingly easy to do. Once a decision-making system like this is place, it is difficult to change, if only because changing to a more qualitative system requires so much more effort. A computer can sort graduate school applicants by their GRE score and even generate the rejection letters to those who miss the cutoff. Making the minimal effort to read the applicants' resumes requires so much more work.

It's more than just laziness that has caused us to stop doing the hard qualitative work in situations like this. There is also a competitive disadvantage to doing things the old-fashioned way. If I spend all my time reading resumes and otherwise digging, by hand, through the records of prospective graduate students, I'll fall behind my colleagues who select their students "by the numbers" and are thus free to spend their time "more productively." I can only justify the labor-intensive approach if it yields measurably (oops, there's that word again) better results. It's a sticky problem.

Some of us, however, are sophisticated enough to use measurement when it suits our purposes, and to ignore it when it doesn't. In the 1980s I attended a seminar presented by the German engineer Helmut Thielsch, an expert on why things go wrong (what we call "failure analysis" in the engineering world). Thielsch, who died in 2003, was something of a rock star in the world of failure analysis. He was particularly good at explaining why things like pipelines and pressure vessels sometimes explode—spectacular stuff like that. He wasn't afraid to get his hands dirty or to go into some rather dangerous situations, and he was also pretty darn smart.

At the seminar, Thielsch was asked if he made use of the science of fracture mechanics in his work. He responded with a yes, noting that one of his Ph.D. employees was an expert in the field. Thielsch added that this expert frequently provided him, Thielsch, with a fracture mechanics–based solution to various problems. "I use those solutions every single time," Thielsch said with a smile, "provided they agree with what I was already planning to do." It struck me at the time that here was a guy who knew how to use measurement—and not be used by it!

There are other examples of this as well. A politician who has the opinion polls—the measured pulse of the electorate—on his side is not afraid to say so. "The American public is overwhelmingly in favor of [insert your favorite popular cause here]. Therefore, I am announcing the following legislative initiative to [further that very same cause]." But what does our political friend do when the polls are not on his side? No problem. "My fellow citizens, we must look not to the polls nor to the pundits to tell us what to do. We must make difficult choices, and do what is right and proper. We must be guided by principles, not polls. Therefore, I am announcing the following legislative initiative . . ." Just like Helmut Thielsch, a politician uses measurement when it suits him, and ignores it when it doesn't.

I could go on with examples like this. The same thing can happen in reverse, as well. We sometimes create measurements simply because they support what we already wanted to do in the first place. Political polls are an obvious example. Politicians often commission slanted polls to provide ammunition for their policies. Sample question from the political Right: "Are you in favor of tax relief for hard-working Americans, or do you favor giving more and more of your hard-earned dollars away to the special interest groups that are sucking the lifeblood out of our country?" Similar question from the Left: "Are you in favor of tax relief for hard-working Americans, or should the poor and middle classes continue to subsidize the lavish lifestyles of the richest Americans—who are sucking the lifeblood out of our country?" Having thus fudged the questions, it's easy for a politician, from either side, to say, "A recent poll shows overwhelming support for my tax policies . . ."

Survey-based measurement results are highly dependant on what questions are asked and how they are asked. In the 2000 U.S. presidential election, for example, the vote among self-reported Catholics went for Al Gore over George W. Bush by 50 to 47%. But those Catholics who responded that they attended church weekly voted for Bush by a 53 to 44% margin. Among voters of all religions, those who reported weekly church attendance voted for Bush by 63 to 36%, a margin of nearly 2 to 1. The key question was not, "What is your religious affiliation?" but rather, "How often do you go to church?"

Other examples of creating measurements to support a cause often include some kind of weighted average. (Beware those weighted averages!) Cost-benefit analysis, for example, is not an inherently evil thing, but, depending on the application, it is so easy to fudge the results that you have to be careful. They are sometimes cited in cases involving some sort of environmental impact, for example to assess a new technology to limit sulfur emissions from coal-burning power plants, or a petroleum pipeline that will run across a protected wilderness area. Cost-benefit analyses in such cases are almost infinitely adjustable, since many of the "costs" and "benefits" are difficult to quantify. If you only look at the bottom line, you get what you deserve. Someone who says, "I ran a cost-benefit analysis, and it says we should go ahead with the project," might only be spouting so much hot air. With weighted averages, I could make that project look as good (or lousy) as I want, just as I could make a butter knife look like the best way to mow your lawn.

Beam Me Up, Scotty!

As the world becomes more and more measured and measurable, there is both a temptation and a driving force to ignore the qualitative aspects of things that are otherwise measurable. We would do well to avoid this temptation, and to work against this driving force.

My favorite television program when I was growing up was *Star Trek* (the original version). The two main characters in *Star Trek* were Captain Kirk and Mr. Spock. With his analytical Vulcan mind, Mr. Spock represented the quantitative nature of things. He could pick apart any problem and provide Captain Kirk with the analytical solution—he could "run the numbers" for the boss. But Kirk's job was harder. First of all, he was in charge—on the starship *Enterprise*, the buck stopped with Captain Kirk. Kirk also had to wrestle with the qualitative side of things. He dealt in emotions and feelings—things Spock's measurements and numbers could never account for. Spock's instruments might tell him that the situation was hopeless—that the enemy Klingon vessel's weapons were unstoppable and its defenses impregnable. But Kirk would somehow win the day. Maybe the Klingon vessel's physical defenses *were* impregnable, but its evil commander was a flawed creature. Kirk would find a way through *his* defenses.

In the beginning of this chapter I spoke of a "vague inner desire for things that can't be measured." That desire is, I believe, somewhat misplaced. Instead, we should embrace measurement in all (well, okay, most) of its myriad forms. At the same time we must be constantly aware of the dual quantitative/qualitative nature of things. We have to be Mr. Spock and Captain Kirk at the same time. To measure is to know? You bet. But the good cook both measures his ingredients and tastes the results.

REFERENCES

CHAPTER 1
Of Love and Luminescence: What, Why, and How Things Get Measured

Berka, K. *Measurement: Its Concepts, Theories and Problems*. London: D. Riedel, 1983.
Gould, Stephen Jay. *The Mismeasure of Man*. New York: Norton, 1981.
Joncich, Geraldine. *The Sane Positivist: A Biography of Edward L. Thorndike*. Middletown, Conn.: Wesleyan University Press, 1968.
Medawar, Peter B. "Unnatural Science." *New York Review of Books*, February 3, 1977, 13–18.
Roberts, Fred S. *Measurement Theory with Applications to Decisionmaking, Utility, and the Social Sciences*. Volume 7 of the *Encyclopedia of Mathematics and Its Application*. Reading, Mass.: Addison-Wesley, 1979.
Thompson, Silvanus P. *The Life of Lord Kelvin*. 1910. 2 vols. Reprint. New York: Chelsea, 1976.
Thomson, Sir William (Lord Kelvin). "Electrical Units of Measurement." Lecture delivered to the Institution of Civil Engineers on May 3, 1883. In *Popular Lectures and Addresses*. London: Macmillan, 1894.
Traub, James. "Harvard Radical." *New York Times Magazine*, August 24, 2003.

CHAPTER 2
Doing the Math: Scales, Standards, and Some Beautiful Measurements

Beranek, L. L. "The Notebooks of Wallace C. Sabine." *Journal of the Acoustical Society of America* 61 (March 1977): 629–39.
Caplow, Theodore, Louis Hicks, and Ben Wattenberg. *The First Measured Century: An Illustrated Guide to Trends in America, 1900–2000*. Washington, D.C.: AEI Press, 2001.

Fowler, Michael. "Galileo and Einstein." Course notes. University of Virginia, 2004.

Hebra, Alex. *Measure for Measure: The Story of Imperial, Metric, and Other Units.* Baltimore: Johns Hopkins University Press, 2003.

Hertzberg, Richard. *Deformation and Fracture of Engineering Materials.* 4th ed. New York: Wiley, 1998.

Nussbaum, Arthur. *A History of the Dollar.* New York: Columbia University Press, 1957.

Roberts, Fred S. *Measurement Theory with Applications to Decisionmaking, Utility, and the Social Sciences.* Volume 7 of the *Encyclopedia of Mathematics and Its Application.* Reading, Mass.: Addison-Wesley, 1979.

Sabine, Wallace C. *Collected Papers on Acoustics.* 1922. Reprint. New York: Dover, 1964.

Sanders, J. H. *Velocity of Light.* New York: Pergamon Press, 1965.

Stevens, S. S. "Measurement and Man." *Science* 127, no. 3295 (February 21, 1958): 383–89.

———. "On the Theory of Scales of Measurement." *Science* 103, no. 2684 (June 7, 1946): 677–80.

———. *Psychophysics: Introduction to Its Perceptual, Neural, and Social Prospects.* Edited by Geraldine Stevens. 1975. Reprint. New Brunswick, N.J.: Transaction Books, 1986.

Woolf, Harry, ed. *Quantification: A History of the Meaning of Measurement in the Natural and Social Sciences.* New York: Bobbs-Merrill, 1961.

CHAPTER 3

The Ratings Game: "Overall" Measurements and Rankings

"America's Best Colleges." *U.S. News and World Report.* 2004.

Ciambrone, David F. *Environmental Life Cycle Analysis.* Boca Raton, Fla.: Lewis, 1997.

Dempsey, Jack, and Barbara Piattelli Dempsey. *Dempsey.* Harper & Row, 1977.

Echikson, William. "Death of a Chef." *New Yorker,* May 12, 2003, 61–67.

Gould, Stephen Jay. *The Mismeasure of Man.* New York: Norton, 1981.

Keeney, Ralph L., and Howard Raiffa. *Decisions with Multiple Objectives.* New York: Cambridge University Press, 1993.

Medawar, Peter B. "Unnatural Science." *New York Review of Books,* February 3, 1977, 13–18.

CHAPTER 4

Measurement in Business: What Gets Measured Gets Done

Cochran, Thomas C. *200 Years of American Business.* New York: Basic Books, 1977.

Daniels, Aubrey C. *Bringing Out the Best in People: How to Apply the Astonishing Power of Positive Reinforcement.* New York: McGraw-Hill, 2000.

Peters, Tom. "What Gets Measured Gets Done." www.tompeters.com, April 28, 1986 (accessed October 3, 2005).

Taylor, Frederick W. *The Principles of Scientific Management.* New York: Harper & Brothers, 1911.

Underhill, Paco. *Why We Buy: The Science of Shopping.* New York: Simon & Schuster, 1999.

CHAPTER 5

Games of Inches: Sports and Measurement

Flynn, James. *Race, IQ, and Jensen.* London: Routledge & Kegan Paul, 1980.

Gleick, James. "'Hot Hands' Phenomenon: A Myth?" *New York Times,* April 19, 1988.

James, Bill. *The Bill James Baseball Abstract, 1988.* New York: Ballantine Books, 1988.

Lewis, Michael. *MoneyBall: The Art of Winning an Unfair Game.* New York: Norton, 2003.

"The Science of Lance Armstrong." http://science.discovery.com/convergence/lance/explore/explore.html (accessed October 3, 2005).

Track-and-field record progression data. www.geocities.com/Colosseum/Arena/3170/index_s.html (accessed August 28, 2005).

Tygiel, Jules. *Past Time: Baseball as History.* New York: Oxford University Press, 2000.

CHAPTER 6

Measuring the Mind: Intelligence, Biology, and Education

"About Functional MRI." Columbia University Functional MRI Research Center. www.fmri.org/fmri.htm (accessed August 28, 2005).

Brooks, David. "Making It: Love and Success at America's Finest Universities." *Weekly Standard,* December 23, 2002.

Caplow, Theodore, Louis Hicks, and Ben Wattenberg. *The First Measured Century: An Illustrated Guide to Trends in America, 1900–2000.* Washington, D.C.: AEI Press, 2001.

College Board. "The New SAT 2005." www.collegeboard.com/newsat/index.html (accessed August 28, 2005).

Eysenck, Hans J. *The Structure and Measurement of Intelligence.* New York: Springer, 1979.

Gardner, Howard. *Frames of Mind: The Theory of Multiple Intelligences.* New York: Basic Books, 1985.

Gould, Stephen Jay. *The Mismeasure of Man.* New York: Norton, 1981.

Katzman, John, Andy Lutz, and Erik Olson. "Would Shakespeare Get into Swarthmore?" *Atlantic Monthly,* March 2004, 97–99.

Kuhn, Thomas S. "Measurement in Modern Physical Science." In *Quantification:*

A History of the Meaning of Measurement in the Natural and Social Sciences, ed. Harry Woolf. New York: Bobbs-Merrill, 1961.

Lemann, Nicholas, *The Big Test: The Secret History of the American Meritocracy*. New York: Farrar, Straus & Giroux, 1999.

Mason, M. S. "Research Starts to Bridge Gap Between Prayer and Medicine." *Christian Science Monitor*, September 15, 1997.

National Research Council. *High Stakes: Testing for Tracking, Promotion, and Graduation*, ed. Jay P. Heubert and Robert M. Hauser. Washington, D.C.: National Academy Press, 1999.

Plato. *The Republic*. Translated and edited by Allan Bloom. New York: Basic Books, 1968, 1991.

Thomson, Sir William (Lord Kelvin). "Electrical Units of Measurement." Lecture delivered to the Institution of Civil Engineers on May 3, 1883. In *Popular Lectures and Addresses*. London: Macmillan, 1894.

U.S. Department of Education. "No Child Left Behind." www.ed.gov/nclb/landing.jhtml (accessed August 28, 2005).

U.S. Supreme Court. *Atkins v. Virginia*, 536 U.S. 304 (2003).

CHAPTER 7

Man: The Measure of All Things

Bilger, Burkhard. "The Height Gap: Why Europeans Are Getting Taller and Taller—and Americans Aren't." *New Yorker*, April 5, 2004, 38–45.

Collins, Francis S., Eric D. Green, Alan E. Guttmacher, and Mark S. Guyer. "A Vision for the Future of Genomics Research." *Nature* 422 (April 24, 2003): 835–47.

Jeffreys, A. J., V. Wilson, and S. L. Thein. "Hypervariable 'Minisatellite' Regions in Human DNA." *Nature* 314 (March 7–13, 1985): 67–73.

Judson, Horace F. *The Eighth Day of Creation: Makers of the Revolution in Biology*. New York: Simon & Schuster, 1979. Reprint. Plainview, N.Y.: Cold Spring Harbor Laboratory Press, 1996.

Ridley, Matt. *Genome: The Autobiography of a Species in 23 Chapters*. New York: HarperCollins, 1999.

Watson, J. D., and F. H. C. Crick. "A Structure for Deoxyribose Nucleic Acid." *Nature* 171 (April 25, 1953): 737–38.

CHAPTER 8

It's Not Just the Heat, It's the Humidity: Global Warming and Environmental Measurement

Burchfield, Joe D. *Lord Kelvin and the Age of the Earth*. New York: Science History Publications, 1975.

Caplow, Theodore, Louis Hicks, and Ben Wattenberg. *The First Measured Century: An Illustrated Guide to Trends in America, 1900–2000*. Washington, D.C.: AEI Press, 2001.

Chamberlain, T. C. "On Lord Kelvin's Address on the Age of the Earth as an Abode Fitted for Life." *Annual Report of the Smithsonian Institution*, 1899, 223–46.

Dalrymple, G. Brent. *The Age of the Earth.* Stanford, Calif.: Stanford University Press, 1991.

Faure, Gunter. *Principles of Isotope Geology.* 2d ed. New York: Wiley, 1986.

Gorst, Martin. *Measuring Eternity: The Search for the Beginning of Time.* New York: Broadway Books, 2001.

Lomborg, Bjorn. *The Skeptical Environmentalist: Measuring the Real State of the World.* New York: Cambridge University Press, 2001.

Mann, Michael E., and Raymond S. Bradley. "Northern Hemisphere Temperatures During the Past Millennium: Inferences, Uncertainties, and Limitations." *Geophysical Research Letters* 26, no. 6 (March 15, 1999): 759–62.

Strahler, Arthur N. *Science and Earth History: The Evolution/Creation Controversy.* Buffalo, N.Y.: Prometheus Books, 1987.

Weart, Spencer. *The Discovery of Global Warming.* Cambridge, Mass.: Harvard University Press, 2003.

CHAPTER 9

Garbage In, Garbage Out: The Computer and Measurement

Clarke, Arthur C. *Profiles of the Future: An Inquiry into the Limits of the Possible.* New York: Harper & Row, 1973.

eBird. Cornell Lab of Ornithology and National Audubon Society. www.ebird.org (accessed August 31, 2005).

Google. www.google.com/technology (accessed October 3, 2005).

Miller, David, et al. *The Cambridge Dictionary of Scientists.* New York: Cambridge University Press, 1996, 2002.

Peterson, Ivars. "Sampling and the Census: Improving the Accuracy of the Decennial Count." *Science News*, October 11, 1997, 238–39.

CHAPTER 10

How Funny Is That? Knowledge Without Measurement?

American Psychiatric Association. *Diagnostic and Statistical Manual of Mental Disorders*, 4th ed., text rev. (DSM-IV-TR). Washington, D.C.: American Psychiatric Association, 2000.

EndocrineWeb.com. "Diagnosing diabetes." www.endocrineweb.com/diabetes/diagnosis.html (accessed October 3, 2005).

Hockney, David. *Secret Knowledge: Rediscovering the Lost Techniques of the Old Masters.* New York: Viking Studio, 2001.

National Oceanic and Atmospheric Administration. www.noaa.gov/tsunamis.html (accessed October 3, 2005).

Stewart, Melissa. "Fingerprints." *Invention and Technology* 17, no. 1 (Summer 2001): 23–30.

U.S. Supreme Court. *Daubert v. Merrell Dow Pharmaceuticals, Inc.*, 509 U.S. 579 (1993).

Woolf, Harry, ed., *Quantification: A History of the Meaning of Measurement in the Natural and Social Sciences.* New York: Bobbs-Merrill, 1961.

CHAPTER 11

Faith, Hope, and Love: The Future of Measurement—and of Knowledge

Cunningham, Marion, and Lauren Jarrett. *The Fannie Farmer Cookbook: Anniversary.* New York: Knopf, 1996.

Edwards, Betty. *Drawing on the Right Side of the Brain: A Course in Enhancing Creativity and Artistic Confidence.* 1979. Rev. ed. Los Angeles: J. P. Tarcher, 1989.

Henshaw, John M. "A Framework and Tools for the Early Decisions in the Product Development Process." Ph.D. diss, University of Delaware, 1990.

Pew Research Center. "Religion and the Presidential Vote: Bush's Gains Broad-Based." December 6, 2004. http://people-press.org/commentary/display .php3?AnalysisID=103 (accessed August 28, 2005).

Reuters. "Happiness is . . . an equation." January 6, 2003. www.cnn.com/2003/ WORLD/europe/01/06/happiness.equation/ (accessed October 3, 2005).

INDEX